NONLINEAR FUNCTIONAL ANALYSIS

Notes on Mathematics and its Applications

E. Artin, ALGEBRAIC NUMBERS AND ALGEBRAIC FUNCTIONS
R. P. Boas, COLLECTED WORKS OF HIDEHIKO YAMABE
M. Davis, A FIRST COURSE IN FUNCTIONAL ANALYSIS
M. Davis, LECTURES ON MODERN MATHEMATICS
J. Eells Jr., SINGULARITIES OF SMOOTH MAPS
K. O. Friedrichs, ADVANCED ORDINARY DIFFERENTIAL EQUATIONS
K. O. Friedrichs, SPECIAL TOPICS IN FLUID DYNAMICS
K. O. Friedrichs and H. N. Shapiro, INTEGRATION IN HILBERT SPACE
A. Guichardet, SPECIAL TOPICS IN TOPOLOGICAL ALGEBRAS
M. Hausner and J. T. Schwartz, LIE GROUPS; LIE ALGEBRAS
P. Hilton, HOMOTOPY THEORY AND DUALITY
F. John, LECTURES ON ADVANCED NUMERICAL ANALYSIS
A. M. Krall, STABILITY TECHNIQUES FOR CONTINUOUS LINEAR SYSTEMS
P. Lelong, PLURISUBHARMONIC FUNCTIONS AND POSITIVE DIFFERENTIAL FORMS
H. Mullish, AN INTRODUCTION TO COMPUTER PROGRAMMING
F. Rellich, PERTURBATION THEORY OF EIGENVALUE PROBLEMS
J. T. Schwartz, W-* ALGEBRAS
J. T. Schwartz, DIFFERENTIAL GEOMETRY AND TOPOLOGY
J. T. Schwartz, NONLINEAR FUNCTIONAL ANALYSIS
J. L. Soule, LINEAR OPERATORS IN HILBERT SPACE
J. J. Stoker, NONLINEAR ELASTICITY

Additional volumes in preparation

Nonlinear Functional Analysis

J. T. SCHWARTZ

Courant Institute of Mathematical Sciences New York University

Notes by

H. Fattorini
R. Nirenberg and H. Porta

with an additional chapter by
Hermann Karcher

GORDON AND BREACH SCIENCE PUBLISHERS
NEW YORK · LONDON · PARIS

150 Fifth Avenue, New York, N.Y.10011

Library of Congress Catalog Card Number: 68-25643

Editorial office for the United Kingdom:

Gordon and Breach Science Publishers Ltd.
12 Bloomsbury Way
London W.C.1.

Editorial office for France:

Gordon & Breach
7-9 rue Emile Dubois
Paris 14e

Distributed in Canada by:

The Ryerson Press
299 Queen Street West
Toronto 2b, Ontario

Editors' Preface

A large number of mathematical books begin as lecture notes; but, since mathematicians are busy, and since the labor required to bring lecture notes up to the level of perfection which authors and the public demand of formally published books is very considerable, it follows that an even larger number of lecture notes make the transition to book form only after great delay or not at all. The present lecture note series aims to fill the resulting gap. It will consist of reprinted lecture notes, edited at least to a satisfactory level of completeness and intelligibility, though not necessarily to the perfection which is expected of a book. In addition to lecture notes, the series will include volumes of collected reprints of journal articles as current developments indicate, and mixed volumes including both notes and reprints.

JACOB T. SCHWARTZ
MAURICE LÉVY

Contents

Introduction

Nonlinear functional analysis is of course not so much a subject, as the complement of another subject, namely, linear functional analysis. In studying our negatively defined field, we will, however, find a certain unity; partly because we shall exclude from functional analysis those analytic theories which do not make use of the characteristic procedure of functional analysis. This characteristic procedure is, of course, the treatment of a given problem, or the construction or study of a desired function, by imbedding the problem or function into a space (generally infinite dimensional) of related problems or functions. In accordance with this distinction we shall, for example, regard much of the asymptotic study (by topological methods) of solutions of nonlinear differential equations as belonging to nonlinear analysis but not to nonlinear functional analysis, while, for instance, the Morse theory of geodesics, or the construction of solutions of partial differential equations by application of the Schauder fixed point theorem, will definitely be considered to belong to nonlinear functional analysis. The distinction suggested is not always clear-cut, however.

We may orient ourselves toward our subject of study as follows. Nonlinear functional analysis is nonlinear analysis in the context of infinite dimensional topological spaces, manifolds, etc. Naturally, our knowledge of nonlinear analysis in this case cannot be more complete than our knowledge of nonlinear analysis in the finite dimensional case. Therefore the finite-dimensional case can serve as a model for the infinite dimensional case. We can formulate our aim as follows: to extend known theorems of nonlinear analysis from the finite to the infinite dimensional case; to analyze any particular difficulties, not present in the finite dimensional case, which arise in the infinite dimensional case.

Now, what are the main branches of nonlinear analysis in finitely many dimensions? They may be listed under five general headings:

1. Elementary calculus.
2. The implicit function theorem and related results.

3. Topological principles for establishing the existence of solutions to systems of equations: the Brouwer fixed point theorem, the theory of degree, the Jordan separation theorem, and, more generally, the Lefschetz fixed point theorem and the general topological intersection theory.

4. Topological theories for establishing the existence of critical points: the Morse critical point theory, and the Lusternik-Schnirelman "category" theory.

5. Theorems following by the powerful special methods of complex function theory.

We shall find infinite dimensional generalizations of theorems belonging to each of these five categories.

1. *Elementary calculus* goes over to B-spaces (and even slightly more general spaces) in a routine way. Integration theory is developed for vector-valued functions defined on a measure space in *Linear Operators*, Chapter 3, and contains no surprises. The proper notion of derivative (as already in two-dimensional spaces) is that of directional derivative or Gateaux derivative, which may be defined as follows. Let ϕ be a function mapping one B-space X into another Y. Then if, for each $x, y \in X$ the function $\phi\,(x + ty)$ of the real variable t is differentiable at $t = 0$, we say that ϕ is (Gateaux) differentiable, and write

$$d\phi\,(x; y) = \frac{d}{dt}\phi\,(x + ty)\bigg|_{t=0}.$$

A certain amount of basically elementary and unsurprising real variable theory is connected with this notion. Thus, for instance, under suitable hypotheses $d\phi\,(x; y)$ is linear in y; so that we may, if we like, speak of the derivative $d\phi(x)$ as a linear operator mapping X into Y. Rather than study this elementary calculus for its own sake, we will develop results which belong to it as needed for other purposes.

2. The *implicit function theorem* in B-spaces exists in two versions. On the one hand, we have the classical "soft" implicit function theorem, which states that if ϕ is a mapping of a B-space X into a B-space Y, if $\phi(0) = 0$, and if ϕ is continuously differentiable and $\phi'(0)$ is a bounded operator with a bounded inverse, then ϕ maps a neighborhood of zero (in X) homeomorphically onto a neighborhood of zero (in Y). This basic version of the theorem has several interesting variants, one of which is the so-called theory of "monotone" mappings. Another class of theorems closely related to this implicit function theorem form the so-called "bifurcation theory". The main

idea of this latter theory may be explained as follows. Suppose that the solutions of a functional equation $\phi(x) = 0$ are to be studied in the vicinity of a given solution $x = 0$. (Here, ϕ is taken to be a differentiable mapping of a B-space X into itself.) We may write $\phi(x) = x + Kx + \psi(x)$, where $|\psi(x)| = 0(|x|^2)$ for x near 0, and where K is a linear transformation. If $(I + K)^{-1}$ exists as a bounded operator, then, by the implicit function theorem, $x = 0$ is an isolated zero. In the bifurcation theory, we suppose only that K is compact, and wish to consider the case in which $(I+K)^{-1}$ does not exist. In this case, it follows by the Riesz theory of compact operators that X decomposes as a direct sum $X = Y \oplus Z$ of two subspaces, both invariant under K, the second being finite dimensional, such that $(I+K)$ is a bounded mapping Y onto itself having a bounded inverse. Correspondingly, we may write $x = [y, z]$, and write the equation $\phi(x) = 0$ as a pair of equations:

$$\phi_1 (y, z) = (I + K) y + \psi_1 (y, z) = 0$$

$$\phi_2 (y, z) = (I + K) z + \psi_2 (y, z) = 0.$$

By the implicit function theorem, the first equation may be solved for y in terms of $z : y = Y(z)$. Substituting this solution into the second equation, we find that the solutions of the original equation $\phi(x) = 0$ are in one-to-one correspondence with the solutions of the equation $(I+K) z + \psi_2(Y(z), z) = 0$. This last equation, however, may be regarded as a finite system of equations in a finite number of variables, upon which all the resources of finite-dimensional analysis may be brought to bear.

An introduction to the theory of bifurcation as outlined above may be found in Graves' article *Remarks on singular points of functional equations*, Trans. Amer. Math. Soc., V. 79, 150–157 (1955).

In addition to the "soft" version of the implicit function theorem described above, there exists in the functional-analytic case a separate "hard" version of the theorem. The precise statement of this second version of the implicit function theorem will be given in a later lecture. At present we shall only remark that this theorem applies even in cases where the Gateaux derivative of ϕ is unbounded as a linear operator, and has an unbounded linear inverse. The theorem is due to J. Nash: *The imbedding problem for Riemannian manifolds*, Ann. Math. 63, pp. 20–63 (1956). J. Moser (*A new technique for the construction of solutions of non-linear differential equations*, Proc. Nat. Acad. U.S.A., V. 47, 1961, pp. 1824–1831) made the useful observation that Nash's "hard" version of the implicit function theorem could be proved by an appropriate modification of the "Newton's Method" of finite dimensional analysis, the superrapid convergence of Newton's method compensating, in an appropriate sense, for the unboundedness of the Frechet derivative and its inverse. Moser has subsequently made interesting applications and extensions of this basic idea, cf. Moser: *On invariant curves of area-preserving mappings of an annulus*, Göttinger Nachrichten, 1962, pp. 1–20, and subsequent publications. Cf. also a lecture by Serge Lang in the 1962 *Séminare Bourbaki.*

3. *Finite codimensional topology.* The attentive reader will have observed that in our listing above of theorems of finite dimensional topology we have separated these theorems into two groups. This separation, somewhat unnatural in the finite dimensional case, is essential in the infinite dimensional case. Consider, for example, the Brouwer fixed point theorem. As is well-known, this theorem is equivalent to the statement that the boundary of the unit sphere is not continuously deformable to a point on itself. In infinite dimensions, however, this statement is false. E.G., if we examine the boundary ∂S of the unit sphere in the Hilbert space $L_2(0, 1)$, and follow the homotopy $f(x) \to f_t(x)$, $1 \geq t \geq \frac{1}{2}$, where

$$\begin{aligned} f_t(x) &= t^{-1/2} f\left(\frac{x}{t}\right), \qquad & 0 \leq x \leq t \\ &= 0 & t \leq x \leq 1 \end{aligned}$$

by the homotopy $f_{1/2}(x) \to t f_{1/2}(x) + \sqrt{1 - t^2}\,\sigma(x)$, $1 \geq t \geq 0$, where $\sigma \in \partial S$ and $\sigma(x) = 0$ for $0 \leq x \leq \frac{1}{2}$, we obtain a continuous deformation of ∂S along itself, to the single point σ. This implies a set of topological consequences rather different from the corresponding results in finite dimensions. We owe to Schauder and Leray the important observation that the most familiar results of finite dimensional topology can be carried over to infinitely many dimensions if attention is restricted to the special category of maps ϕ having the form $\phi = I + \psi$, where I is the identity, and ψ is a mapping whose range is compact. Thus, for instance, if we confine our attention to this special category of maps, the boundary of the unit sphere is not continuously deformable to a single point along itself. Moreover, again for maps of this category, a straightforward generalization of the finite dimensional theory of degree can be established, and infinite dimensional generalizations of many of the basic theorems of finite dimensional topology obtained. As a basic reference, see Schauder and Leray: *Topologie et équations fonctionelles*, Ann. Sci. Ecole Norm. Sup. (3) 51 (1934) pp. 45–78. The infinite dimensional theory of degree is based upon the finite dimensional method, which we shall develop by a simplified method patterned after the procedure of Heinz: *An elementary analytic theory of degree in n-dimensional space*, J. Math. Mech. 8 (1959) pp. 231–247.

An especially useful theorem belonging to this circle of ideas is the fixed point theorem of Schauder: any continuous mapping into itself of a compact convex set in a locally convex linear topological space possesses a fixed point. Cf. Schauder: *Der Fixpunktsatz in Funktionalräumen*, Studia Math. 2 (1936) pp. 171–180. Krein and Rutman (Uspekhi Math. Nauk 3, No. 1, pp. 3–95) give an interesting application of fixed point theory to the "projective space" of a *B*-space.

A connected account of many of the principal results in the type of functional topology discussed above is given by A. Granas: *The theory of compact vector fields and some of its applications to topology of functional spaces* (I). Roszprawy Math. XXX, Warsaw 1962. Granas lays stress on the homotopy theory of compact maps and on the Borsuk antipodal point theorem, but avoids the theory of degree.

4. *Finite dimensional topology*. The second category of topological results available in functional spaces is distinguished by the fact that it makes reference to the ordinary singular homology and cohomology groups, defined similarly in the functional case and in the finite dimensional case. The Morse theory and the Lusternik-Schnirelman theory both begin with the same construction. A *manifold* is defined to be a topological space locally homeomorphic to a given *B*-space in such a way that the "transition mappings" between the various "local coordinate patches" which cover the manifold are infinitely often differentiable. On such a manifold, all the ordinary local notions of analysis such as directional derivative, differentiable function, etc., are available. Let M be such a manifold, and let f be a smooth real-valued function defined on M. A *critical point* of f is by definition a point in M at which the directional derivative of f in every direction vanishes. If M has a Riemannian metric, we may in the usual way define the gradient ∇f of f, which is a field of vectors tangent to M; in this case, the critical points of f are the points p where $\nabla f(p) = 0$. If there exists no critical point p of f such that $a \leq f(p) \leq b$ (and assuming certain additional, technical hypotheses), then the subsets $M_a = \{q \in M | f(q) \leq a\}$ and $M_b = \{q \in M | f(q) \leq b\}$ are diffeomorphic. To see this, we have only to note that if each point q such that $a \leq f(q) \leq b$ is pushed down in the direction of the gradient field ∇f until M_a is reached, we obtain the desired diffeomorphism. This statement is the first main lemma of the Morse theory. The second observation on which the Morse theory is built gives a corresponding result for the case in which $\{q \in M | a \leq f(q) \leq b\}$ contains an isolated set of critical points. In this case, and under the further assumption that the critical points are all nondegenerate

in an appropriate sense, a closer analysis shows that the space M_b is diffeomorphic to a space $M_a \cup H$ obtained from M_a by affixing a certain collection of "handles". Thus the sequence of critical points of f describes the construction of M by the successive addition of "handles" to a "ball". This connection may be exploited in either of two directions: to conclude from the known topology of M that any function f defined on M must admit critical points of certain numbers and types, or, conversely, to deduce information about the topology of M from a knowledge of the critical points of some particular function on M.

If we let M be the space of all smooth curves on a finite dimensional manifold N, and regard M in an appropriate way as being an infinite dimensional manifold, then the general Morse theory outlined above reduces to the special Morse theory of geodesics.

A lucid account of the Morse theory, especially in the finite-dimensional case, is to be found in Milnor: *Morse theory*, Ann. of Math., Study 51, Princeton 1963. The generalization to infinite dimensional manifolds is developed in Palais: *Lectures on Morse theory*, Notes, Harvard, 1963, to be republished in 1964 in the Journal *Topology*. Palais gives the application of the general theory to the Morse geodesic theory, developing in detail an account of the necessary compactness properties of the infinite dimensional manifold M and the function f on it. Further applications of the general theory to establish the existence of higher type critical structures in theories of minimal surfaces, etc., are to be hoped for. We may also refer to a set of notes entitled *Lectures of Smale on Differential Topology* (Columbia, 1963). These notes give extensions of various qualitative theorems of finite-dimensional differential topology to the infinite dimensional case.

The Lusternik-Schnirelman theory of critical points agrees with the Morse theory in making use of the deformations along gradient curves on a manifold M. However, the methods of Lusternik-Schnirelman are more point set theoretic than those of Morse, and lead to more general but less precise results. If A and M are topological spaces, and ϕ maps A into M and is continuous, call ϕ a map *of category* 1 if it is homotopic to a constant map, and call ϕ a map *of category* k if A can be divided into k sets $A_1, \dots, A_k$, but no fewer, such that $\phi | A_i$ is of category 1. If $A \subseteq M$, the category cat (A) is the category of the identity map of A into M. It is not hard to establish that cat $(A) - 1$ is a lower bound for the topological dimension of A. If f is a real valued function defined on A, and $m \leq$ cat (A), put

$$c_m(f) = \inf_{\operatorname{cat}(B) \geq m} [\sup \{f(x) | x \in B\}].$$

Then $c_1(f) \leq c_2(f) \leq \cdots$. It may be shown, under suitable compactness hypotheses, that for each $m \leq \text{cat}(A)$ there exists a set $B_m \subseteq A$ such that $\text{cat}(B_m) = m$, and $\sup\{f(x) \mid x \in B_m\} = c_m(f)$. Were it the case that A contained no critical point q with $f(q) = c_m(f)$, we could push all the points $p \in B_m$ down in the direction of the gradient field ∇f, obtaining a set $\bar{B}_m$ of category m such that $\sup\{f(x) \mid x \in B_m\} < c_m(f)$, a contradiction. Thus we see that each value $c_m(f)$ must be a critical value of f. A refinement of this argument shows that if $c_m(f) = c_n(f)$, then $\{x \in A \mid f(x) = c_m \text{ and } \nabla f(x) = 0\}$ must be of category at least $m - n + 1$. Thus any smooth function on A must admit at least cat (A) critical points. This last result makes it important to be able to establish lower bounds for the category of a space. We will see in a subsequent lecture that such results follow from an analysis of the singular cohomology ring of a topological space. For an introductory account of the theory of category and some of its applications, cf. Lusternik and Schnirelman, *Méthodes topologiques dans les problèmes variationels*, Gauthier-Villars, Paris, 1934.

Chapter VIII of the present notes, generously contributed by Dr. Hermann Karcher*, gives an account of some of the Morse Theory of closed geodesics on manifolds which are topological spheres, according to methods stemming from Klingenberg.

5. *The complex analytic case.* A few results applying specifically to complex analytic functional mappings between complex linear spaces are known. In the first place, one has the usual elementary results guaranteeing the power series expansion of complex analytic mappings, etc. The bifurcation theory, where applicable, shows that the set of zeroes of an analytic functional equation $\phi(x) = 0$ is in one-to-one bianalytic correspondence with the set of zeroes of a similar set of analytic equations in a finite number of complex variables. A good deal is known about the structure of such *analytic varieties*, and the bifurcation theory enables one to carry all this information over to the functional case.

It follows readily from the definition of degree, in the cases where this definition is applicable, that the degree of a complex analytic map $x \to \phi(x)$ near any isolated zero is non-negative. According to an interesting theorem of Jane Cronin (cf. Cronin: *Analytic Functional Mappings*, Ann. Math. 58

* The work of H. Karcher was supported at the Courant Institute of Mathematical Sciences, New York University, by the National Science Foundation under Grand NSF-GP-8114.

(1953) pp. 175–181) the degree of such a zero is actually positive. This result, combined with the results available from the general theory of degree, leads to a principle of permanance of zeroes that generalizes the well-known theorem of Rouché to the functional case.

6. *Miscellany:* In addition to the five principal categories of results outlined above, a variety of miscellaneous special results must be included in our subject. These will be noted as they arise in our subsequent lectures. In the present introduction, we shall note the work of Hammerstein (cf. *Nichtlineare Integralgleichungen nebst Anwendungen,* Acta Math., V. 54 (1930) pp. 117–176) on integral equations, in which the order properties of the integral operators studied are exploited. This work is related to the theory of monotone operators alluded to above.

We may also mention the existence of various investigations, notably those of E. Rothe, devoted to the *variational method* in functional analysis, i.e., to the possibility of solving functional equations $\phi(x) = 0$ by casting them into the form $\Phi(x) = \min$, where Φ is an appropriately selected functional.

While the literature on the subject of the present course of lectures is somewhat scattered, a number of useful books have appeared. We mention in the first place the book of Krasnoselskii: *Topological methods in the theory of non-linear integral equations,* Moscow, 1956, 392 pp. An English translation of a related survey article by Krasnoselskii appears in the AMS Translations, Ser. 2, No. 10, pp. 345–409. Krasnoselskii gives a good account of the available information on continuity and compactness of nonlinear integral operators of various forms, a good summary account of a number of other important topics in nonlinear theory, as well as an extensive bibliography. A less closely related, but still relevant work is the article *Functional analysis and applied mathematics* by Kantorovic in Uspekhi Math. Nauk 3 (No. 6) (1948), pp. 89–185, as well as this author's treatise *Approximated methods of higher analysis.* An account of the differential calculus in B-spaces is to be found in the book of Michal: *Le calcul différentielle dans les espaces de Banach* (V. I, Fonctions analytiques—équations intégrales) Gauthier-Villars, 1958 (150 pp.), in the well-known treatise by Hille and Phillips on semigroups, and in the texts of advanced calculus by Dieudonne and by Serge Lang.

The reader wishing to extend his knowledge of nonlinear functional analysis beyond the necessarily limited material contained in the present notes will find it useful to consult the comprehensive survey article by James Eells: *A Setting for Global Analysis,* Bull. Amer. Math. Soc. v. 72, 1966, p. 751–809. This excellent review may also serve as a guide to the literature of the subject.

CHAPTER I

Basic Calculus

A. Some Definitions and a Lemma on Topological Linear Spaces

1.1. Definition: We say that E is a *topological linear space* iff E is a linear space which is given a topology such that addition and multiplication by scalars are continuous functions, i.e.: $+ : E \times E \to E$ and $\cdot : E \times R \to E$ are continuous functions, where $E \times E$ and $E \times R$ have the product topology.

1.2. Definition: Let E be a T.L.S. We say that E is *locally convex* iff there exists a family of convex sets $\{U\}$ which is a basis for the family of neighborhoods of 0.

(A set K is called *convex* iff $x, y \in K$ implies $tx + (1 - t)y \in K$ for every $t \in [0, 1]$.)

1.3. Definition: A T.L.S. will be called an *F-space*, or *Frechet space*, iff, as a topological space, it is metric and complete, with a topology given by a "norm" function $|x|$ which satisfies: (i) $|x|$ real $\geqq 0$; (ii) $|x| = 0$ iff $x = 0$; (iii) $|x + y| \leqq |x| + |y|$. (See *Linear Operators**, Chapter 2.) We shall write *L.C.F.-space* for locally convex F-spaces.

**Linear Operators,* Nelson Durnford and Jacob T. Schwartz, Wiley-Interscience, Vol. 1, 1958, Vol. 11, 1963.

1.4. Definition: A T.L.S. will be called a *Banach space*, or a *B-space*, iff it is complete and its topology is given by a norm, which in addition to conditions (i), (ii) and (iii) of the above definition, satisfies (iv) $|\lambda x| = |\lambda|\,|x|$.

The spaces with which we shall ordinarily deal are L.C.F.-spaces.

1.5. Definition: Let E be an F-space. We say that $K \subset F$ is *bounded* iff for any neighborhood U of 0 there exists $\varepsilon \to 0$ such that $\varepsilon K \subset U$.

This condition is easily seen to be equivalent to the following one: $\varepsilon_n \to 0$ and $k_n \in K$ implies $\varepsilon_n k_n \to 0$.

We shall now prove a lemma relating L.C.F.-spaces to B-spaces to B-spaces.

1.6. Lemma: A L.C.F.-space is a B-space if it contains a bounded open set.

First, we note that boundedness is unaffected by translations, so we can assume that $0 \in U$, where U is bounded and open. By definition of an L.C.F.-space, U will contain a convex neighborhood U' of 0, which a fortiori is bounded. Now, we can replace U' by $V = U' \cap (-U')$ which is also convex, bounded, and a symmetric neighborhood of 0, i.e., $V = -V$. By the definition of a bounded set, the family $\{\varepsilon V\}$, ε real > 0, is a neighborhood basis at 0. We consider now the support function of V, $p(x)$, defined by $p(x) = \left[\sup_{tx \in V} |t|\right]^{-1}$. (Obviously if in a B-space V is the unit sphere, $p(x) = |x|$.) The function $p(x)$ has the four properties of a norm function:

(i) $p(x)$ real and $\geqq 0$. Obvious. It is a finite number because V is absorbing.

(ii) $p(x) = 0 \Leftrightarrow x = 0$. If $p(x) = 0$, $\sup_{tx \in V} |t| = \infty$, and this means that $tx \in V$ for all t, because V is convex. Hence $x \in \dfrac{1}{t} V$ for all $t > 0$, whence x is in every neighborhood of 0, since $\left\{\dfrac{1}{t} V\right\}$ is a neighborhood basis. Therefore $x = 0$, because the space is Hausdorff.

(iii) $p(x + y) \leqq p(x) + p(y)$. It is apparent that $p(x)$ can be defined by $p(x) = \inf_{x \in tV,\, t>0} t$. Let $\alpha, \beta > 0$ and such that $x \in \alpha V$ and $y \in \beta V$. Then $x + y \in \alpha V + \beta V$; since V is convex, $\alpha V + \beta V = (\alpha + \beta) V$, whence $x + y \in (\alpha + \beta) V$. Therefore $\inf_{t>0,\, x+y \in tV} t \leqq \inf_{t>0,\, x \in tV} t + \inf_{t>0,\, y \in tV} t$, which is (iii).

(iv) $p(\alpha x) = |\alpha|\, p(x)$. If $\alpha > 0$, it is easy to see that $p(\alpha x) = \alpha p(x)$. But since V is symmetric, (iv) holds for any α, since $\alpha V = -\alpha V$.

Next we note that $V = \{x | p(x) < 1\}$ if we assume V to be open. For then clearly $x \in V$ implies $p(x) < 1$. Also $p(x) < 1$ implies $x \in tV$ for some $t < 1$, i.e., $x = tv$, $v \in V$; since V is convex, $x \in V$. We see at once then that $\varepsilon V = \{x | p(x) < \varepsilon\}$. Therefore $p(x)$ is continuous at 0, and consequently at every x.

Conversely, for any $\varepsilon > 0$, there exists $\delta > 0$ such that $p(x) < \delta$ implies $|x| < \varepsilon$. Simply choose δ so that $\delta V \subset S_\varepsilon$, where $S_\varepsilon = \{x | |x| < \varepsilon\}$. This is possible since $\{\varepsilon V\}$ is a neighborhood basis at 0.

Thus we have shown that $| \ |$ and $p(x)$ determine the same topology; hence $p(x)$ is the required norm.

Q.E.D.

B. Elementary Calculus

1.7. Definition: Let X and Y be T.L.S. Let U be an open subset of X and $f: U \to Y$. We say that f has a *Gateaux derivative* $df(x, y)$ at $x \in U$ iff

$$\frac{d}{dt} f(x + ty)\bigg|_{t=0} = df(x, y)$$

exists for every $y \in X$.

We call this derivative the *derivative of f at x in the direction y*, and shall write it often as $(df(x))(y)$ or $(f'(x))(y)$.

1.8. Definition: Let X and Y be T.L.S. and let $\phi : U \to Y$, where U is a neighborhood of 0 in X. We say that ϕ is horizontal at 0 iff for each neighborhood V of 0 in Y there exists a neighborhood U' of 0 in X, and a function $0(t)$ such that

$$\phi(tU') \subset o(t) V.$$

1.9. Definition: Let X, Y be T.L.S. and U open in X. Let $f: U \to Y$ and $x_0 \in U$. We say that f is *Frechet differentiable*, or *F-differentiable at* x_0, iff there exists a continuous linear map $A : X \to Y$ such that if we write

$$f(x_0 + y) = f(x_0) + Ay + \phi(y),$$

then ϕ is horizontal at 0.

We call A the derivative of f at x_0, and we write it $df(x, y)$ as in Definition 1.7.

1.10. Remark: If the spaces are B-spaces, then the definition of a function horizontal at 0 is equivalent to

$$|\phi(x)| \leqq |x| \, \psi(x)$$

where ψ is real valued and $\lim_{x \to 0} \psi(x) = 0$. Thus, in a B-space, the condition of F-differentiability can be expressed as follows:

$$f(x_0 + y) = f(x_0) + Ay + 0(|y|).$$

1.11. Remark: If a linear function λ is horizontal at 0, then $\lambda = 0$, as follows at once from the definition. Thus we see that the F-derivative of a function is unique, because if

$$f(x_0 + y) = f(x_0) + Ay + \phi(y)$$

and

$$f(x_0 + y) = f(x_0) + By + \phi'(y)$$

where A and B are continuous and linear, and ϕ and ϕ' are horizontal at 0, then $A - B$ is horizontal at 0 (the sum of ϕ and ϕ' is still horizontal at 0) whence $A = B$.

1.12. Remark: The domain of f in the definition of G-differentiability can be assumed to be a "finitely open set in x", where x is simply a linear space. Also, for complex spaces, it is easy to see that the Gateaux derivative is always linear in y, and that the hypothesis of linearity is also unnecessary for F-derivatives. (Cf. Hille and Philips [1], Sections 3.13 and 26.3.)

In the case X is a B-space, it is easy to show that F-differentiability implies Gateaux differentiability.

1.13. Lemma: Let f: $U \to Y$, where U is open in a B-space X and Y is a T.L.S. Then if f has an F-derivative at x_0, it also has a Gateaux derivative at x_0, and they are equal.

Proof: We write

$$f(x_0 + ty) - f(x_0) = tAy + o(|ty|),$$

where A is linear and continuous. But $o(|ty|) = o(|t|\,|y|)$ and

$$\lim_{t \to 0} \frac{1}{t} (f(x_0 + ty) - f(x_0)) = Ay.$$

Q.E.D.

The next lemma gives the chain rule for F-derivatives.

1.14. Lemma: If f: $U \to V$ is F-differentiable at x_0, and $g: V \to W$ is F-differentiable at $f(x_0)$, then $g(f(x))$ is F-differentiable at x_0, and its derivative is given by:

$$(d(gf))(x_0, y) = dg(f(x_0), df(x_0, y)).$$

Here, U, V and W are open sets contained in X, Y, Z which are T.L.S.

Proof: We have only to write:

$$
\begin{aligned}
g\,[f(x_0 + y)] &= g\,[f(x_0) + df(x_0, y) + \phi(y)] \\
&= g\,[f(x_0)] + dg\,[f(x_0), df(x_0, y)] \\
&\qquad + dg\,[f(x_0), \phi(y)] + \psi\,[df(x_0, y) + \phi(y)],
\end{aligned}
$$

where ϕ and ψ are horizontal at 0. It is easy to see that the last term is horizontal at 0 as a function from X to Z. We note that if ϕ is horizontal at 0, and if λ is linear and continuous, then $\lambda \circ \phi$ is also horizontal at 0. This follows immediately from the definition of horizontality. Thus $dg\,[f(x_0), \phi(y)]$ is horizontal at x_0, and its derivative is $dg\,[f(x_0), df(x_0, y)]$.

Q.E.D.

We next prove another lemma relating Gateaux and F-differentiability.

1.15. Lemma: Let X and Y be B-spaces, U open in X and $f\colon U \to Y$. If f has a Gateaux derivative $f'(x, y)$ in U, which is linear in the variable y, and if, when regarded as a linear operator, $f'(x)$ is bounded for $x \in U$ and depends continuously on x in the uniform topology, then f is F-differentiable en U.

Proof: Our point of departure is the formula

$$
\frac{d}{dt} f(x + ty) = f'(x + ty)\,(y)
$$

which one can prove easily. It follows that

$$
\begin{aligned}
f(x + y) &= f(x) + \int_0^1 f'(x + ty)\,(y)\,dt \\
&= f(x) + f'(x)\,(y) + \int_0^1 [f'(x + ty) - f'(x)]\,(y)\,dt.
\end{aligned}
$$

But now:

$$
\begin{aligned}
\left| \int_0^1 [f'(x + ty) - f'(x)]\,(y)\,dt \right| &\leqq |y| \int_0^1 |f'(x + ty) - f'(x)|\,dt \\
&= |y|\,o(1) = o(|y|).
\end{aligned}
$$

Q.E.D.

1.16. Remark: In the last lemma we integrated functions of a real (or complex) variable with values in a B-space (cf., for example, Hille and Phillips [1], Chapter III). The basic fact we used was that

$$
\left| \int_S f(S)\,d\mu\,(S) \right| \leqq \int_S |f(S)|\,d\mu\,(S).
$$

This is not true in general for F-spaces, because its proof depends upon the inequality $|\Sigma \alpha_i f_i| \leqq \Sigma |\alpha_i| \, |f_i|$. In L.C.F. spaces, however, one can define weak integration by appropriate use of linear functionals. (Cf. loc. cit.) Local convexity implies separation theorems which assure the uniqueness of the integral.

1.17. Lemma: *(Contracting Mapping Principle.)* Let X be a complete metric space and $\phi : U \to X$, U open in X, and assume $\varrho(\phi(x), \phi(y)) \leqq \alpha \varrho(x,y)$ with $0 \leqq \alpha < 1$, where $\varrho(x, y)$ is the distance between x and y. Moreover, suppose there exists $z_0 \in U$ such that $\varrho(z_0, X - U) > M$, and $\varrho(z_0, \phi(z_0)) < M(1 - \alpha)$. Then there exists a fixed point $z_\infty = \phi(z_\infty)$ such that $\varrho(z_0, z_\infty) < M$.

Proof: $\varrho(z_0, \phi(z_0)) < M(1 - \alpha) < M < \varrho(z_0, X - U)$, so $\phi(z_0)$ is also in U, and inductively $\phi^2(z_0), \ldots, \phi^n(z_0) \ldots$ are all in U, where $\phi^n(z_0) = \phi(\phi^{n-1}(z_0))$. The sequence $z_0, \phi(z_0), \ldots, \phi^n(z_0) \ldots$ is Cauchy, as follows from the contracting hypothesis. Hence we can set $z_\infty = \lim_n \phi^n(z_0)$. By the continuity of ϕ, $\phi(z_\infty) = z_\infty$. The formula

$$\varrho(z_0, \phi^n(z_0)) < M(1 - \alpha^n)$$

is easily proved by induction on n. Then

$$\varrho(z_0, z_\infty) = \varrho(z_0, \lim \phi^n(z_0)) = \lim \varrho(z_0, \phi^n(z_0)) < M.$$

Q.E.D.

C. The "soft" Implicit Function Theorem

1.18. Lemma: Let x be an F-space, U the sphere $\{x : |x| < r\}$, and $\phi : U \to X$ such that $\phi(x) = x + \psi(x)$, where $\psi(x)$ satisfies:

$$|\psi(x) - \psi(y)| \leqq \alpha |x - y| \quad \text{with} \quad 0 \leqq \alpha < 1, \quad \text{and} \quad \psi(0) = 0.$$

Then: (i) $\phi(U)$ covers a sphere of radius $r(1 - \alpha)$ about 0. (ii) ϕ is one-to-one and the inverse ϕ^{-1} satisfies a Lipschitz condition with constant $1/1 - \alpha$.

Proof: (i) We apply the last lemma to the function $f(x) = -\psi(x) + p$ where $p \in X$ and $|p| < r(1 - \alpha)$. If we put $z_0 = 0$, this inequality implies that $|z_0 - \phi(z_0)| = |p| < r(1 - \alpha)$. Hence there exists a point z_∞ in U such that $z_\infty = -\psi(z_\infty) + p$, i.e., $\phi(z_\infty) = p$. (ii) Suppose $\phi(x) = x + \psi(x) = p$

and $\phi(y) = y + \psi(y) = q$. Then, $x - y + \psi(x) - \psi(y) = p - q$, and $|x - y| = |\psi(x) - \psi(y)| \leqq |p - q|$, so $(1 - \alpha)\,|x - y| \leqq |p - q|$, and we are done.
Q.E.D.

1.19. Corollary: If x is a B-space and if, in the notation of the above lemma, $\psi'(x)$ exists and $|\psi'(x)| \leqq \alpha < 1$ in U, and $\psi(0) = 0$, then (i) and (ii) are true.

Proof: We only have to note that

$$|\psi(x) - \psi(y)| \leqq \int_0^1 |\psi'(x + t(x - y))|\,|x - y|\,dt \leqq \alpha\,|x - y|.$$

Q.E.D.

Now we can prove the following important theorem:

1.20. Theorem: (Implicit function theorem.) Let X, Y be B-spaces and $\phi: U \to Y$, where U is an open neighborhood of 0 in X and $\phi(\alpha) = 0$. Assume: (a) ϕ is F-differentiable in U. (b) $\phi'(x)$ depends continuously on x in the uniform operator topology. (c) $\phi'(0)$ is a bounded linear map with a bounded linear inverse. Then ϕ maps a sufficiently small neighborhood of zero homeomorphically onto a neighborhood of zero.

Proof: Let $A = \phi'(0)$. We put $\eta = A^{-1} \circ \phi$. Then $\eta: U \to X$, η has an F-derivative $\eta'(x)$ which is continuous in x in the uniform operator topology, and $\eta'(0) = I$, the identity operator. Let $\psi = \eta - I$. Then $\psi'(0) = 0$, and

$$\psi(x) - \psi(y) = \eta(x) - \eta(y) - (x - y) = \int_0^1 \eta'(x + t(x - y))(x - y)\,dt$$

$$- (x - y) = \int_0^1 [\eta'(x + t(x - y)) - I)(x - y)\,dt.$$

Thus

$$|\psi(x) - \psi(y)| \leqq |x - y| \int_0^1 |\eta'(x + t(x - y)) - I|\,dt.$$

But we can make the integral on the right of the last formula less than one, by taking x and y in a sufficiently small neighborhood $V \subset U$ of 0. Then the preceding lemma applies to η, and, a fortiori, $\phi = A\eta$ maps V homeomorphically onto a neighborhood of 0 in Y.
Q.E.D.

1.21. Corollary: Given the conditions of 1.20, the inverse map $\phi^{-1}: \phi(V) \to V$ is F-differentiable. Setting $\psi = \phi^{-1}$, we have, for $y \in \phi(V)$, the formula

$$\psi'(y) = (\phi'(\phi^{-1}(y)))^{-1}.$$

Proof: If $\phi(x_1) = y_1$, $\phi(x_2) = y_2$, then

$$\begin{aligned}&|\phi^{-1}(y_2) - \phi^{-1}(y_1) - (\phi'(x_1))^{-1}(y_2 - y_1)| \\ &\quad = |(\phi'(x_1))^{-1}[\phi'(x_1)(x_2 - x_1) - (y_2 - y_1)]| \\ &\quad \leqq A\,|\phi'(x_1)(x_2 - x_1) - \phi(x_2) + \phi(x_1)|.\end{aligned}$$

This last expression is $o(|x_2 - x_1|)$, whence the first expression is $o(|y_2 - y_1|)$, and the result follows.

Q.E.D.

An induction argument easily yields the fact that if ϕ has derivatives of higher order (definition given later), then so does ϕ^{-1}. Similarly, if ϕ depends continuously on some parameter, so does ϕ^{-1}.

The following theorem is a global version of the "local" implicit theorem:

1.22. Theorem: Let X and Y be B-spaces, and $\phi : X \to Y$ a continuously F-differentiable function, and suppose ϕ' is invertible (as a linear operator) at every $x \in X$, and moreover, that $\|[\phi'(x)]^{-1}| \leqq K < \infty$ uniformly in x. Then ϕ is a homeomorphism of X *onto* Y.

The proof will depend on the following lemma:

1.23: Lemma: Under the same hypothesis as the theorem, if Δ is the square $0 \leqq s \leqq 1$, $0 \leqq t \leqq 1$, and if $F(s, t)$ satisfies the conditions:

(i) $F(s, t)$: $\Delta \to Y$.

(ii) $F(s, t)$ is continuous in (s, t) and for every fixed s, $0 \leqq s \leqq 1$, $F(s, t)$ is F-differentiable in t.

(iii) $F(s, t)$ has fixed endpoints, i.e. there exist $y_0, y_1 \in Y$ such that $F(s, 0) = y_0$, $F(s, 1) = y_1$ for all $0 \leqq s \leqq 1$.

Then there exists a function $G(s, t)$ from Δ to X which also satisfies (ii) and in addition

$$\phi\,[G(s, t)] = F(s, t) \quad \text{for all} \quad (s, t) \in \Delta.$$

Proof of the lemma: By the local implicit function theorem, there exist neighborhoods V of y_0 and U of x_0 (where $\phi(x_0) = y_0$) such that ϕ is a homeomorphism of U onto V. Then, for sufficiently small ε, we can define $G(s, t)$ as $\phi^{-1}(F(s, t))$ if $0 \leqq t \leqq \varepsilon$ and for $0 \leqq s \leqq 1$. We call a the largest of the values such that $G(s, t)$ can be defined in the rectangle $0 \leqq t < a$, $0 \leqq s \leqq 1$. Assume $a < 1$. If $G(s, t)$ is defined for $t = a$, consider the curve $G(s, a)$ and its image $\phi(G(s, a)) = F(s, a)$. For each s, $0 \leqq s \leqq 1$, we can select a neighborhood U_s of $G(s, a)$ and a neighborhood V_s of $F(s, a)$

such that ϕ is a homeomorphism of U_s onto V_s. But $G(s, a)$, $0 \leqq s \leqq 1$ is compact, and therefore, there exists a finite subcovering of the curve $G(s, a)$ with neighborhoods U_{s_i}, $i = 1, \ldots, n$. In each of these neighborhoods, we can define the function $G(s, t)$ for all s and $0 \leqq t < a + \varepsilon_n$, by the local implicit function theorem. So, $G(s, t)$ can be defined for the rectangle $0 \leqq s \leqq 1$, $0 \leqq t < a + \min_n \varepsilon_n$, contradicting the fact that a was the largest of such numbers. Now $G(s, t)$ must be F-differentiable in t, for $F(s, t)$ satisfies (ii) and ϕ^{-1} is locally F-differentiable. By the chain rule, we have for all $0 \leqq s \leqq 1$:

$$\phi' [(G(s, t)] G'(s, t) = F'(s, t)$$

where the prime denotes differentiation with respect to t. So:

$$G'(s, t) = [\phi'(G(s, t))]^{-1} F'(s, t)$$

and

$$|G'(s, t)| \leqq |[\phi'(G(s, t))]^{-1}| \, |F'(s, t)| \quad \text{for all} \quad 0 \leqq s \leqq 1.$$

Then

$$|G'(s, t)| \leqq K|F'(s, t)| \leqq A_s.$$

Now, integrating with respect to t between t_0 and t_1, we get:

$$|G(s, t_1) - G(s, t_0)| \leqq A_s |t_1 - t_0|,$$

a Lipschitz condition for $G(s, t)$. Therefore $\lim_{t \to a-} G(s, t)$ exists for each s, and $G(s, t)$ can be defined at $t = a$. We have proved that $a = 1$.
Q.E.D.

Proof of the theorem: Let $y_0 = \phi(0)$, and y any point in Y. Consider the straight line segment joining y_0 and y; we write it $y(t)$, $0 \leqq t \leqq 1$. As a particular case of the lemma, there exists a curve $x(t)$, $0 \leqq t \leqq 1$, in X such that $\phi[x(t)] = y(t)$. Then $\phi[x(1)] = y$, and ϕ is onto.

Let, as before, $y_0 = \phi(0)$, and suppose there are two points $x_0, x_1 \in X$, $x_0 \neq x_1$ such that $\phi(x_0) = \phi(x_1) = y$. We take two curves $x_0(t)$ and $x_1(t)$ joining respectively 0 to x_0 and 0 to x_1, with $0 \leqq t \leqq 1$. Then both image curves $y_0(t) = \phi(x_0(t))$ and $y_1(t) = \phi(x_1(t))$ will join y_0 and y. As Y is simply connected, there exists a function $F(s, t)$ from Δ to Y continuous in (s, t) and such that $F(0, t) = y_0(t)$, $F(1, t) = y_1(t)$, and $F(s, 0) = y_0$, $F(s, 1) = y$.

By the argument of the lemma, we find a function $G(s, t)$ from Δ to X, continuous and such that $\phi(G(s, t)) = F(s, t)$, with $G(0, t) = x_0(t)$ and $G(1, t) = x_1(t)$. But then the continuous curve $G(s, 1)$ with endpoint x_0 and x_1 is mapped by ϕ onto y. This contradicts the local implicit function theorem, and therefore ϕ is $1 - 1$. ϕ^{-1} is obviously continuous (it is F-differentiable) so ϕ is a homeomorphism of X onto Y.

Q.E.D.

D. The Hilbert Space Case

Now we shall prove some implicit function results for Hilbert space:

1.24. Lemma: Let H be a real Hilbert space, L a bounded linear operator mapping H into H. Suppose that for every $x \in H$, $(Lx, x) \geqq a(x, x)$ where $a > 0$. Then L^{-1} is defined everywhere, and $|L^{-1}| \leqq a^{-1}$.

Proof: L is $1 - 1$, for if $Lx = 0$, then $0 = (Lx, x) \geqq a(x, x)$, and so $x = 0$. The range of L is closed, for if $x_n \in$ range of L and $x_n \to x$, then $L^{-1}x_n$ is a Cauchy sequence:

$$(L^{-1}x_n - L^{-1}x_m, L^{-1}x_n - L^{-1}x_m) \leqq \frac{1}{a^2}(x_n - x_m, x_n - x_m)$$

$$= \frac{1}{a^2}|x_n - x_m|^2.$$

Therefore $L^{-1}x_n \to y \in H$; since L is continuous, $Ly = x$. Now let $z \in H$ be in the orthogonal complement of the range of L. Then for every $x \in H$, $(z, Lx) = 0$ implies $(z, Lz) = 0$ implies $(z, z) = 0$, and $z = 0$. We have proved that the range of $L = H$, and L is onto. From the inequality $(Lx, x) \geqq a(x,x)$ we see that $|Lx| \geqq a$. Hence $|L^{-1}x| \leqq a^{-1}$.

Q.E.D.

1.25. Corollary: Let $\phi : H \to H$ where H is a Hilbert space, and suppose ϕ is continuously F-differentiable, and $(\phi'(x)y, y) \geqq a(y, y)$ for every x and y. Then ϕ is a homeomorphism of H onto H.

Proof: Apply the last lemma and Theorem 1.22.

Q.E.D.

We shall state the hypothesis of this corollary in a slightly different way.

1.26. Definition: We say that $\phi : H \to H$ (H is a Hilbert space) is *strongly monotone* if for every $x, y \in H$ we have

$$(\phi(x) - \phi(y), x - y) \geqq a(x - y, x - y)$$

for some $a > 0$.

It is easy to see that a differentiable ϕ is strongly monotone if $(\phi'(x)y, y) \geqq a(y, y)$ for every $x, y \in H$. In fact, suppose ϕ is strongly monotone. Then for any real t:

$$(\phi(y + tz) - \phi(y), z) \geqq ta|z|^2, \quad \text{where} \quad y, z \in H,$$

and dividing by t and taking the limit, we get $(\phi'(y)z, z) \geqq a(z, z)$. Conversely, we get the condition of strong monotonicity by integrating the condition involving the derivative.

The following definition will be useful in the sequel:

1.27. Definition: $\phi : H \to H$ will be called *monotone* if for every $x, y \in H$, $(\phi(x) - \phi(y), x - y) \geqq 0$. If the sign $>$ holds for $x - y \neq 0, \phi$ will be called *strictly monotone.*

1.28. Remark: Obviously strongly monotone implies strictly monotone implies monotone. Furthermore, ϕ is strongly monotone with a constant a if $\frac{1}{a}\phi - I$ is monotone, and if ϕ is strictly monotone (a fortiori, if ϕ is strongly monotone), ϕ is $1 - 1$.

We prove now a useful lemma on Euclidean space:

1.29. Lemma: (Kirszbraun) Suppose $\{x_1 \cdots x_n\}$ and $\{x_1' \cdots x_n'\}$ are two sets of points in E^n, and let p be also in E^n. Assume that for every i, j, $1 \leqq i, j \leqq n$, we have $|x_i' - x_j'| \leqq |x_i - x_j|$ $\left(| \ | \text{ is the standard norm } \sqrt{\Sigma\, |x_n^2|}\right)$. Then, there exists $p' \in E^n$ such that

$$|p' - x_j'| \leqq |p - x_j| \quad \text{for every } j, \quad 1 \leqq j \leqq n.$$

Proof: Let

$$\lambda = \inf_{p' \in E^n} \max_{1 \leqq i \leqq n} \frac{|p' - x_i'|}{|p - x_i|}.$$

This infimum is assumed at some point $p^+ \in E^n$, for $\max_{1 \leqq i \leqq n} \frac{|p' - x_i'|}{|p - x_i|}$ becomes large when p' is large. Hence we can put

$$\max_{1 \leqq i \leqq n} \frac{|p^+ - x_i'|}{|p - x_i|} = \lambda.$$

Now, suppose that for $1 \leqq i \leqq k$ we have $|p^+ - x_i'| = \lambda\,|p - x_i|$, and that for $k + 1 \leqq i \leqq n$ we have $|p^+ - x_i'| < \lambda\,|p - x_i|$. We shall show that $p^+ \in \overline{co}\,(x_1', \ldots, x_k')$ (the convex hull of $\{x_1', \ldots, x_k'\}$). Suppose that $p^+ \notin \overline{co}\,(x_1' \cdots x_k')$. Then, we can separate p^+ from $\overline{co}\,(x_1' \cdots x_k')$ with a hyperplane Λ. If we move the point p^+ toward Λ perpendicularly, it is obvious that the distance from p^+ to every point in the halfspace not containing p^+

decreases. We can move p^+ by so little as to preserve the inequalities $|p^+ - x_i'| < \lambda |p - x_i|$, $k+1 \leqq i \leqq n$, and now we get $|p^+ - x_i'| < \lambda |p - x_i'|$ for every $1 \leqq i \leqq n$, which is impossible, because p^+ realizes the infimum. Thus we can express p^+ as $\sum_1^k c_i x_i'$, where $c_i \geqq 0$ and $\sum_1^k c_i = 1$. Let $R_i = p - x_i$ and $R_i' = p^+ - x_i'$. Now suppose λ is greater than 1. Then

$$R_i'^2 > R_i^2 \quad \text{for} \quad i = 1 \cdots k. \tag{1}$$

On the other hand, we have by the hypothesis:

$$(R_i' - R_j')^2 \leqq (R_i - R_j)^2,$$

and after expanding and using (1):

$$R_i' R_j' > R_i R_j, \quad 1 \leqq i, j \leqq k. \tag{2}$$

Now, as $\sum_1^k c_i = 1$, we have $\left(\sum_1^k c_i\right) p^+ = \sum_1^k c_i x_i'$, and therefore, $\sum_1^k c_i R_i' = 0$, and, by (1) and (2), $0 > (\Sigma\, c_i R_i)^2$, a contradiction. We have proved that $\lambda \leqq 1$.

Q.E.D.

Now, it is easy to generalize this result for Hilbert space:

1.30. Corollary: Let $\{x_\alpha\}$ and $\{x_\alpha'\}$ be two sets of points in the Hilbert space H, and $p \in H$. Suppose $|x_\alpha' - x_\beta'| \leqq |x_\alpha - x_\beta|$. Then there exists $p' \in H$ such that $|x_\alpha' - p'| \leqq |x_\alpha - p|$ for all α.

Proof: We want prove that the intersection of the infinite family of spheres with center x_α' and radius $|x_\alpha - p|$ is non-void. But spheres are compact in the weak topology for H, so it is sufficient to prove that every finite subfamily of spheres has a non-void intersection. If, then, there are only finitely many x_α's, the set $\{x_\alpha\} \cup \{x_\alpha'\}$ generates a finite dimensional Euclidean space, and we have only to apply the lemma.

Q.E.D.

1.30A. As the following counterexample (due to Charles McCarthy) shows, the obvious generalization of Kirszbraun's lemma to Banach spaces that are not Hilbert spaces is not true in general. We give the following

Theorem: Let l_n^p, $1 < p < \infty$, $n \geqq 1$ be n-dimensional Euclidean space with the norm

$$|x|_p = |(x_1 \cdots x_n)|_p = \left(\sum_{i=1}^n |x_i|^p\right)^{1/p}.$$

Then if $n > 1$, $p \neq 2$, the generalization of Kirszbraun's lemma does not hold.

Proof: Take in l_n^p, $p > 2$, $n > 1$ the points $x_1 = (0, 0, \ldots, 0)$, $x_2 = (0, 1, 0, \ldots, 0)$, $x_3 = (1, 0, \ldots, 0)$. Evidently

$$|x_1 - x_2|_p = |x_1 - x_3|_p = 1, \quad |x_2 - x_3| = 2^{1/p}.$$

Choose now spheres S_1, S_2, S_3 around x_1, x_2, x_3 of radii $2^{(1-p)/p}$. We have

$$S_1 \cap S_2 \cap S_3 = \{(\tfrac{1}{2}, \tfrac{1}{2}, 0, \ldots, 0)\}.$$

Now let

$$x_1' = (0, 0, \ldots, 0), \quad x_2' = ((1 - 2^{1-p})^{1/p}, 2^{(1-p)/p}, 0, \ldots, 0),$$

$$x_3' = ((1 - 2^{1-p})^{1/p}, -2^{(1-p)/p}, 0, \ldots, 0).$$

Again

$$|x_1' - x_2'|_p = |x_1' - x_3'|_p = 1, \quad |x_2' - x_3'| = 2^{1/p}.$$

But if we take spheres S_1', S_2', S_3' of radii $2^{(1-p)/p}$ around x_1', x_2', x_3', their intersection will be void. In fact, by uniform convexity

$$S_2' \cap S_3' = \{((1 - 2^{1-p})^{1/p}, 0, \ldots, 0)\} = \{y\}.$$

Since $(1 - 2^{1-p})^{1/p} > 2^{(1-p)/p}$ is $p > 2$, $S_1' \cap S_2' \cap S_3' = \phi$. The case l_n^p, $1 < p < 2$, $n > 1$ may be handled in a similar way; the points x_1, x_2, x_3 are replaced by x_1', x_2', x_3' and vice versa, and the radii of S_1, S_2, S_3 become $(1 - 2^{1-p})^{1-p}, 2^{(1-p)/p}, 2^{(1-p)/p}$ respectively.

Q.E.D.

1.31. Theorem: Let H be a Hilbert space, S any subset of H, and $\phi : S \to H$. Suppose $|\phi(x) - \phi(y)| < K\,|x - y|$ for all $x, y \in S$. Then ϕ can be extended to all of H in such a way that the extension satisfies the same Lipschitz condition.

Proof: Without loss of generality we can suppose that $K = 1$. By Zorn's lemma, there exists a maximal extension $\bar{\phi}$ subject to the same Lipschitz condition. Suppose $p \notin$ domain of $\bar{\phi}$. We have $|\bar{\phi}(x) - \bar{\phi}(y)| \leqq |x - y|$ for $x, y \in$ domain ϕ. Therefore, by the last corollary, we can find $p' \in H$ such that

$$|\bar{\phi}(x) - p'| \leqq |x - p| \quad \text{for all} \quad x \in \text{domain}\ \bar{\phi}.$$

If we define $p' = \phi(p)$, we have extended $\bar{\phi}$ to one more point preserving the Lipschitz condition, and thus contradicting the maximality of $\bar{\phi}$. Hence domain of $\bar{\phi} = H$.

Q.E.D.

We now make some definitions preparatory for the next theorem.

1.32. Definition: We say that $\phi : X \to Y$ (X, Y are B-spaces), is *feebly continuous* iff the mapping $t \to \phi(x + ty)$ is continuous from R to Y with the weak topology for every pair $x, y \in X$.

1.33. Definition: $\phi : X \to Y$ (X, Y are B-spaces) is *slightly continuous* iff $x_\alpha \to x$ strongly in X implies $\phi(x_\alpha) \to \phi(x)$ weakly in Y.

1.34. Remark: As it is easily seen, continuity implies slight continuity implies feeble continuity.

1.35. Theorem (Minty): (a) Let $\phi : S \to H$ (H is a Hilbert space), be defined in an open set $S \subset H$, and suppose ϕ is feebly continuous and strongly monotone. Then ϕ is an open mapping. (b) Let $\phi : H \to H$ be defined everywhere, and suppose ϕ is slightly continuous and strongly monotone. Then ϕ maps H *onto* H.

Proof: (a) As we remarked earlier, we can assume without loss of generality that $\phi = id + T$, where T is monotone. Now consider the Hilbert direct sum $H \oplus H$. We introduce the relations:

(1) $$[x, y]\, M\, [x', y'] \quad \text{iff} \quad (x - x', y - y') \geqq 0$$

and

(2) $$[x, y]\, L\, [x', y'] \quad \text{iff} \quad |y - y'| \leqq |x - x'|.$$

(Note that neither is transitive.)

Let $\phi : H \oplus H \to H \oplus H$ (Cayley transformation) be defined by

(3) $$\Phi([x, y]) = \frac{1}{\sqrt{2}}[x + y, x - y].$$

It is easy to see that Φ is an isometry (of course $|[x, y]|^2 = |x|^2 + |y|^2$), and that $\Phi^2 = id$.

Now let $p = [x, y]$ and $q = [x', y']$. We have:

(4) $$pMq \quad \text{iff} \quad \Phi(p)\, L \Phi(q).$$

For

$$\begin{aligned}
&\Phi(p)\, L\Phi(q)\\
&\quad \text{iff} \quad |x - y - x' + y'|^2 \leqq |x + y - x' - y'|^2\\
&\quad \text{iff} \quad -2(x - x', y - y') \leqq 2(x - x', y - y')\\
&\quad \text{iff} \quad (x - x', y - y') \geqq 0 \quad \text{iff} \quad pMq.
\end{aligned}$$

Call $\Gamma \subset H \oplus H$ the graph of T. Since T is monotone,

$$pMq \quad \text{for all} \quad p, q \in \Gamma. \tag{5}$$

By (4), if we put $\Gamma_1 = \Phi(\Gamma)$, we get:

$$pLq \quad \text{for all} \quad p, q \in \Gamma_1. \tag{6}$$

This means that Γ_1 is the graph of some function S_1 satisfying a Lipschitz condition with $K = 1$. Obviously the domain of S_1 is the set of points $\frac{1}{\sqrt{2}}(x + y) \in H$ such that $[x, y] \in \Gamma$, i.e.,

$$\text{domain}\,(S_1) = \frac{1}{\sqrt{2}}\,\text{range}\,(id + T). \tag{7}$$

By the previous theorem, we can extend S_1 to a function S_2 defined on all of H and satisfying the same Lipschitz condition. Let Γ_2 = graph of S_2 and $\Gamma_3 = \Phi(\Gamma_2)$. Then $\Gamma_3 \supset \Gamma$, because $\Gamma_2 \supset \Gamma_1$. S_2 satisfies the Lipschitz condition, whence

$$p, q \in \Gamma_2 \quad \text{implies} \quad pLq \tag{8}$$

and

$$p, q \in \Gamma_3 \quad \text{implies} \quad pMq \tag{9}$$

(apply (4) and recall that $\Phi^2 = id$).

Now, by (3) and (7) we have:

$$\Gamma = \left\{\frac{1}{\sqrt{2}}\,[(id + S_2)\,x, (id - S_2)\,x]; \quad x \in \frac{1}{\sqrt{2}}\,\text{range}\,(id + T)\right\} \tag{10}$$

and

$$\Gamma = \left\{\frac{1}{\sqrt{2}}\,[(id + S_2)\,x, (id - S_2)\,x]; \quad x \in H\right\}. \tag{11}$$

Suppose now that the range of $id + T$ = range of ϕ is not open; then there exists a point in this range which is a limit of a sequence of points not in the range, and by (10) and (11), this means that there exists a point $[y, z] \in \Gamma$ such that $[y, z] = \lim_n [y_n, z_n]$, and $\{[y_n, z_n]\} \subset \Gamma_3 - \Gamma$. But, by hypothesis, the domain of T is open; hence for some n_0, $y_{n_0} \in$ domain of T, because $y \in$ domain of T and $\lim y_n = y$. Putting $y^* = y_{n_0}$, $z^* = z_{n_0}$, we arrive at the following conclusion: There exists a pair $[y^*, z^*]$ such that:

(i) $y^* \in$ domain of T,

(ii) $[y, z]\, M\, [y^*, z^*]$ for every pair $[y, z]\, \Gamma \in \Gamma$,

(iii) $z^* \neq Ty^*$.

Now we show that this leads to a contradiction.

By (i), for small $\varepsilon > 0$, $y = y^* + \varepsilon\,(z^* - Ty^*)$ belongs to the domain of T, so, by (ii), we have:

$$(y^* - y, z^* - Ty) \geqq 0,$$

and using the definition of y:

$$-\varepsilon\,(z^* - Ty^*, z^* - T\,(y^* + \varepsilon\,(z^* - Ty^*))) \geqq 0$$

or

$$(z^* - Ty^*, z^* - T\,(y^* + \varepsilon\,(z^* - Ty^*))) \leqq 0.$$

As $\varepsilon \to 0$, we have, using the feeble continuity of T:

$$|z^* - Ty^*|^2 \leqq 0.$$

Thus, $z^* = Ty^*$, which contradicts (iii). This proves that range of ϕ is open.

(b) As we remarked before, ϕ is $1 - 1$, because it is strongly monotone, and moreover ϕ^{-1} satisfies a Lipschitz condition with constant $1/a$, as follows from Definition 1.26 and Schwarz's inequality. Now to prove that ϕ is onto we use the same argument as was used in Theorem 1.22, in proving that if $x(t)$ is defined for $t < a$, it is defined for $t = a$; as before, we use, the Lipschitz condition on ϕ^{-1} to prove that $x(t)$ has a limit as $t \to a$. Then we use the slight continuity of ϕ to prove that $\phi\,(x(a)) = y(a)$.

Q.E.D.

We now establish an additional theorem for monotone functions:

1.36. Theorem: Suppose $\phi : H \to H$ (H a Hilbert space) is monotone and continuous. If $p \in H$ is such that $(x, \phi(x) - p) \geqq 0$ for $|x| \geqq R_p$ (where R_p is a number depending on p), then p belongs to the range of ϕ.

Proof: We can assume that $p = 0$. Let $\phi_\varepsilon(x) = \varepsilon x + \phi(x)$ with $\varepsilon > 0$. Then ϕ_ε is strongly monotone, and by Minty's theorem there exists $x_\varepsilon \in H$ such that

(1) $$\varepsilon x_\varepsilon + \phi(x_\varepsilon) = 0,$$

So that multiplying by x_ε we have

$$\varepsilon|x_\varepsilon|^2 + (\phi(x_\varepsilon), x_\varepsilon) = 0.$$

Therefore, for every $\varepsilon > 0$ $|x_\varepsilon|$ must be smaller than R_p, so that the x_ε form a bounded set. Hence there exists a sequence $\{\varepsilon_n\} \to 0$ such that the sequence $\{x_{\varepsilon_n}\}$ tends weakly to a point x_∞, and we can suppose that $|x_{\varepsilon_n}|$ also converges. Now, by the monotonicity of ϕ and (1), we get:

$$(x_\delta - x_\varepsilon, \delta x_\delta - \varepsilon x_\varepsilon) \leqq 0 \quad \text{for every} \quad \delta > 0 \quad \text{and} \quad \varepsilon > 0.$$

Then, if we put $\delta = \varepsilon_n$ and let $n \to \infty$,

$$(x_\infty - x_\varepsilon, -\varepsilon x_\varepsilon) \leqq 0 \quad \text{for every} \quad \varepsilon > 0$$

or

$$(x_\infty - x_\varepsilon, x_\varepsilon) \geqq 0.$$

Therefore

(2) $$|x_\infty|^2 \geqq \lim_{n\to\infty} |x_{\varepsilon_n}|^2.$$

On the other hand, spheres in B-spaces are weakly closed, and this means that:

(3) $$|x_\infty| \leqq \lim |x_{\varepsilon_n}|.$$

Hence, by (2) and (3)

$$|x_\infty| = \lim |x_{\varepsilon_n}|$$

and so

$$x_{\varepsilon_n} \to x_\infty \text{ strongly}.$$

By the continuity of ϕ we get $\phi(x_\infty) = 0$.
Q.E.D.

1.37. Example: Suppose (S, Σ, μ) is a finite measure space, and V a finite dimensional linear space. Let f: $S \times V \to V$ be continuous in V for every S, and such that $|f(s, u)| \leqq K|u| + 1$ for some constant $K \geqq 1$ and every $s \in S$, $u \in V$. Then $\phi(s) \to f(s, \phi(s))$ maps $L^2(S, V)$ into $L^2(S, V)$ because

$$\int_S |f(s, \phi(s))|^2 d\mu \leqq K^2 \int_S (|\phi(s)| + 1)^2 d\mu.$$

We call this map $\phi(F)$. If a sequence $\{\phi_n(s)\}$ converges in L^2 to $\phi(s)$, then there exists a subsequence $\{\phi_{n_i}(s)\}$ converging to $\phi(s)$ a.e. Hence $F(\phi_{n_i})(s) \to F(\phi)(s)$ a.e. But $|F(\phi_{n_i})(s)| \leqq K|\phi_{n_i}(s)| + 1$, whence there exists a subsequence $\{\phi_{n_{i_j}}\}$ such that $\left|\phi_{n_{i_j}}(s)\right| \leqq \psi(s)$ for all j, where $\psi(s)$ is a summable function. (Simply take $\{\phi_{n_{i_j}}\}$ such that $\sum_{j=1}^{\infty} \|\phi_{n_{i_j}} - \phi\| < \infty$.) By the Lebesgue theorem, $F(\phi_{n_{i_j}}) \to F(\phi)$ in L_2. This proves that F is continuous.

Now let D be a linear operator in L_2 with bounded inverse, and suppose we want to solve the equation

$$D^*FDx = y.$$

By Minty's theorem, if D^*FD is strongly monotone then there exists a solution for any $y \in L_2$. The condition for strong monotonicity is

$$(D^*FDx_1 - D^*FDx_2, x_1 - x_2) \geqq \varepsilon\,|x_1 - x_2|^2 \quad \text{where } \varepsilon > 0, \quad x_1, x_2 \in L_2,$$

i.e.,

$$(Fx'_1 - Fx'_2, x'_1 - x'_2) \geqq \varepsilon'\,|x'_1 - x'_2|^2 \quad \text{where } \varepsilon' > 0, \quad x'_1, x'_2 \in L_2.$$

(Calling $x'_i = Dx_i$, $i = 1, 2$, and remembering that D is bounded and has a bounded inverse.)

We therefore see that for solvability it is sufficient to have

$$(f(s, v) - f(s, v'), v - v') \geqq \varepsilon'\,|v - v'|^2 \quad \text{for all} \quad s \in S, \quad v, v' \in V.$$

(Note that here the scalar product is that in R^n, whereas before it was the one in L_2.)

This condition is implied by:

$$(df(s, v)\,v', v') \geqq \varepsilon'\,|v'|^2 \quad \text{for all} \quad s \in S, \quad v, v' \in V$$

which is equivalent to the following condition: There exists an $\varepsilon > 0$ such that the symmetric matrix ${}^tJ + J - \varepsilon I$ is positive definite, where J is the Jacobian matrix of f. Thus for the existence of solutions at every point of the above equation, it is sufficient to require that the matrix $A = \left\{\left(\dfrac{\partial f^i}{\partial x_j} + \dfrac{\partial f^j}{\partial x_i}\right)_{i,j}\right\}$ has smallest eigenvalue >0 at every point.

E. Compact Mappings

1.38. Definition: Let E, F be two T.L.S. $\phi : E \to F$ is *compact* iff it is continuous and maps bounded sets into compact sets, i.e., if $B \subset E$ is bounded, then $\phi(B)$ is relatively compact.

1.39. Definition: $\phi : E \to F$ is called *locally compact* at a point $p \in E$ iff ϕ is continuous in a neighborhood V of p, and maps V into a relatively compact set.

1.40. Theorem: Let E, F be T.L.S., F complete. Let $\phi: E \to F$ be F-differentiable at $p \in E$ and locally compact at p. Then $d\phi(p)$ is a compact linear operator.

Proof: We may suppose that $p = 0$ and that $\phi(p) = 0$. Let $A = d\phi(p)$ and suppose A is not compact. Let S be a bounded subset of E, with non-compact $\overline{A(S)}$. By the completeness of F, we can find a family $\{x_\alpha\} \subset A(S)$ and a neighborhood U of O in F such that $x_\alpha - x_\beta \notin U$ whenever $a \neq \beta$. Now let $\{y_\alpha\} \subset S$ be such that $Ay_\alpha = x_\alpha$, and for any $\delta > 0$, let us define:

$$\eta_\delta(x_n) = \phi(\delta y_n).$$

Then we have:

$$(1) \qquad \eta_\delta(x_n) - \delta x_n = \phi(\delta y_n) - \delta x_n = A(\delta y_n) - \delta x_n + \psi(\delta y_n) = \psi(\delta y_n)$$

where ψ is a function horizontal at 0, i.e. for every neighborhood V of I in F, there exists a neighborhood U of 0 in E such that

$$\psi(\delta U) \subset o(\delta)\, V.$$

Now

$$(2) \qquad \eta_\delta(x_\alpha) - \eta_\delta(x_\beta) = (\delta x_\alpha - \delta x_\beta) + (\eta_\delta(x_\alpha) - \delta x_\alpha) + (\delta x_\beta - \eta_\delta(x_\beta)).$$

Choose a symmetric and circled neighborhood V of 0 in F such that $V + V + V \subset U$. For such a neighborhood there exists another one W in E such that

$$\psi(\delta W) \subset o(\delta)\, V.$$

Since S is bounded, there exists $\lambda > 0$ such that $\lambda S \subset W$. Then for every α,

$$\eta_\delta(x_\alpha) - \delta x_\alpha = \psi(\delta y_\alpha) \in \psi\left(\frac{\delta}{\lambda} W\right) \subset o\left(\frac{\delta}{\lambda}\right) V.$$

But as $\dfrac{1}{\delta}\, o\left(\dfrac{\delta}{\lambda}\right) \to 0$ as $\delta \to 0$, it follows that for sufficiently small δ

$$o\left(\frac{\delta}{\lambda}\right) V \subset \delta V.$$

Therefore, by (2), $\eta_\delta(x_\alpha) - \eta_\delta(x_\beta) \notin \delta V$ whenever $\alpha \neq \beta$, because if $\eta_\delta(x_\alpha) - \eta_\delta(x_\alpha) \in \delta V$, then $\delta x_\alpha - \delta x_\beta + \delta V + \delta V \subset \delta U$ contrary to our assumption. Hence $\phi(\delta y_\alpha) - \phi(\delta y_\beta) \notin \delta U$ for $\alpha \neq \beta$, and for sufficiently small δ, this contradicts the local compactness of ϕ.

Q.E.D.

F. Higher Differentials and Taylor's Theorem

We recall that if $X_1, X_2, \ldots, X_n, Z$ are linear spaces over the same scalar field, a function M: $(X_1 \times X_2 \times \cdots \times X_n) \to Z$ is multilinear or n-linear iff it is linear in each of the variables separately. If $X_1, \ldots, X_n, Z$ are B-spaces M is continuous iff there exists a constant K such that $|M(x_1 \cdots x_n)| \leqq K\,|x_1|\,|x_2| \cdots |x_n|$ for all x_j in X_j, i.e. if it is *bounded.* The minimum of the numbers K satisfying this inequality will be called the *norm,* of M, $|M|$. The set of all n-linear bounded maps M from $X_1 \times X_2 \times \cdots \times X_n$ to Z will be denoted by $B(X_1, \ldots, X_n; Z)$, and it is easy to verify that if X_j and Z are Z are B-spaces then $B(X_1, \ldots, X_n; Z)$ is a B-space with the usual addition and scalar multiplication, and with the norm defined above. In the case $X_1 = X_2 = \cdots = X_n$, $B(X_1, \ldots, X_n; Z)$ will be written $B^n(X, Z)$.

1.41. Lemma: Let $X_1, \ldots, X_n, Z$ be B-spaces over the same scalar field. Then there is an isometric isomorphism between $B(X_1, \ldots, X_n; Z)$ and $B(X_1, B(X_2, \ldots, B(X_n, Z)) \ldots)$.

The proof is left as an exercise for the reader.

Suppose that f is F-differentiable on a set D_0 in X with range of f in Z. Then the function f_1, defined for $x \in D_0$ by $f_1(x) = df(x)$, has its values in the B-space $B(X, Z)$. It makes sense to ask if f_1 is differentiable. If it is, then the differential of $f_1 = df$ will have its values in the space $B^2(X, Z)$ of bounded bilinear functions of X to Z, where, by the lemma above, we have identified $B^2(X, Z)$ and $B(X, B(X, Z))$. We define the differential of $f_1 = df$ at a point c to be the second differential of f at c and we denote this second differential by $d^2f(x)$. Hence $d^2f(c)$ is a bounded bilinear function on X to Z. Higher order differentials are defined by induction.

1.42. Definition: A function f on $D \subseteq X$ to Z is said to be *in class C^n on D*, written $f \in C^n$, iff the n-th differential d^nf exists at every point of D and the mapping $x \to d^nf(x)$ of D into $B^n(X; Z)$ is continuous. If $f \in C^n$ for all n, we say that $f \in C^\infty$. Observe that if $f\colon X \to Y \in C^n$ on a neighborhood of a point $c \in X$ and if $g\colon Y \to Z \in C^n$ on a neighborhood of the point $b = f(c)$, then $h = g \circ f\colon X \to Z \in C^n$ on a neighborhood of c.

We now prove Taylor's theorem for B-spaces. We shall write $x^{(k)}$ for the k-tuple $(x, x, \ldots, x)$.

1.43. Theorem: Suppose that $f \in C^n$ on an open set D which contains the line segment joining c to $c + x$. Then

$$f(c + x) = f(c) + \frac{1}{1!}\, df(c; x) + \frac{1}{2!}\, d^2f(c; x^{(2)}) + \cdots$$
$$+ \frac{1}{(n-1)!}\, d^{n-1}f(c; x^{(n-1)})$$
$$+ \frac{1}{n!}\int_0^1 (1 - t)^{n-1}\, d^nf(c + tx; x^{(n)})\, dt.$$

Proof: Since the map $t \to d^nf(c + tx;\ x^{(n)})$ is continuous on $[0, 1]$ to Z, it is clear that both sides of the equation have a meaning. To establish the equality let $\mathbf{Z^*}$ be a continuous linear functional on Z and let F be defined on $[0, 1]$ to the scalar field by $F(t) = Z^*f(c + tx)$. Then $F^{(k)}(t) = Z^*\,[d^kf(c + tx; x^{(k)})]$ for $0 \leqq k \leqq n$ and we can apply the scalar-valued form of Taylor's theorem to F. If we observe that Z^* commutes with integration, we can apply the Hahn-Banach theorem and obtain the result. Q.E.D.

1.44. Corollary: Under the hypotheses of Taylor's theorem, there exists a bounded n-linear function R_n from X to Z such that

$$f(c + x) = f(c) + \frac{1}{1!}\, df(c; x) + \cdots + \frac{1}{(n-1)!}\, d^{n-1}f(c; x^{(n-1)}) + R_n(x^{(n)}).$$

Proof: Let

$$R_n = \frac{1}{n!}\int_0^1 (1 - t)^{n-1}\, d^nf(c + tx)\, dt.$$

Q.E.D.

1.45. Corollary: Under the hypotheses of Taylor's theorem, there exists a function Q on a neighborhood of the origin in X to Z such that

$$f(c + x) = f(c) + \frac{1}{1!}\, df(c; x) + \cdots + \frac{1}{(n-1)!}\, d^{n-1}f(c; x^{(n-1)})$$
$$+ \frac{1}{n!}\, d^nf(c; x^{(n)}) + Q(x),$$

where $Q(x) = o(|x|^n)$.

Proof: Observe that

$$\frac{1}{n}\int^1 (1 - t)^{n-1}\, dt = 1,$$

and define $Q(x)$ to be

$$Q(x) = \frac{1}{(n-1)!}\int_0^1 (1-t)^{n-1}\,[d^n f(c+tx; x^{(n)}) - d^n f(c; x^{(n)})]\,dt.$$

Q.E.D.

Note: The reader may easily generalize the definitions and results of this section to the case of locally convex T.L.S.

G. Complex Analyticity

Let X and Y be complex B-spaces.

1.46. Definition: We say that $\phi: X \to Y$ is *complex analytic* on an open subset 0 of X iff $\phi(z_1x_1 + z_2x_2 + \cdots + z_kx_k)$ is analytic in $z_1, \ldots, z_k$ for every $x_1, \ldots, x_k$ in D, z_i complex.

If ϕ is complex analytic, we have immediately

$$\phi(x) = \frac{1}{2\pi i}\oint \frac{\phi(x+\zeta y)}{\zeta}\,d\zeta \qquad \text{(Cauchy's formula)}$$

and

$$d^n\phi(x; y, \ldots, y) = \frac{n!}{2\pi i}\oint \frac{\phi(x+\zeta y)}{\zeta^{n+1}}\,d\zeta.$$

1.47. Lemma: If $v(x+iy)$ is a c^1 vector-valued function of x and y, and if $\frac{1}{i}\frac{\partial v}{\partial x} = \frac{\partial v}{\partial y}$, then v is complex analytic.

Proof: Let v^* be a continuous linear functional. $v^*(v(x+iy))$ is a complex valued function which satisfies the Cauchy-Riemann equations and is therefore analytic. Thus

$$v^*(v(z)) = \frac{1}{2\pi i}\oint v^* \frac{v(\zeta)}{\zeta - z}\,d\zeta,$$

and by the Hahn-Banach theorem, the Cauchy formula holds, and v is analytic.

Q.E.D.

Suppose ϕ is analytic in D. Assume that $0 \in D$ and $\phi'(0)$ is invertible. By the implicit function theorem, ϕ has a local inverse, ψ. If u and $v + 2u$ are sufficiently near 0,

$$\phi(\psi(u)) = u \quad\text{and}\quad \phi(\psi(v+2u)) = v + 2u.$$

Then

$$\phi'(\psi(v))\frac{d}{dt}\psi(v+tu)\Big|_{t=0} = u$$

and

$$\phi'(\psi(v))\frac{d}{dt}\psi(v+itu)\Big|_{t=0} = iu.$$

Since $\phi'(\psi(v))$ is invertible,

$$\frac{d}{dt}\psi(v+itu) = i\frac{d}{dt}\psi(v+tu).$$

By the preceding lemma, $\psi(v+zu)$ is analytic in z, and consequently ψ is analytic. We have therefore proved the implicit function theorem in the analytic case:

1.48. Theorem: Under the hypotheses of the implicit function theorem 1.20, and if ϕ is analytic, ϕ^{-1} is also complex analytic.

H. Derivatives of Quadratic Forms

1.49. Definition: If B, V are linear spaces and $f: B \to V$, we say that f is a *quadratic form* iff the expression

(1) $$\beta(x, y) = f(x+y) - f(x) - f(y)$$

is bilinear and

(2) $$f(\lambda x) = \lambda^2 f(x)$$

for every $x \in B$ and every scalar λ.

It is clear that in such a case $f(-x) = f(x)$ and $f(0) = 0$. β is called the *bilinear form associated* with f. From the definition it follows that

(3) $$f(x) = \tfrac{1}{2}\beta(x, x),$$

and therefore f and β determine each other. Suppose now that B and V are Banach spaces. It is clear that f is continuous if and only if β is continuous. If β is continuous, then

(4) $$|f(x)| \leqq \tfrac{1}{2}\|\beta\|\,|x|^2,$$

where $\|\beta\|$ stands for the norm of β as a bilinear function. If we define $\|f\| = \frac{1}{2}\|\beta\|$, (4) may be written

$$|f(x)| \leqq \|f\|\,|x|^2.$$

The main theorem on derivatives of quadratic forms is the following:

1.50. Theorem: Every continuous quadratic form has F-derivatives of all orders. Denote by β the bilinear form associated with f and by $\|f\|$ the number $\frac{1}{2}\sup|\beta(x, y)|$, $|x| \leqq 1$, $|y| \leqq 1$. The first and second derivatives of f are:

$$f'(z)\,h = \beta(z, h),$$

$$f''(z) = \beta,$$

and the higher derivatives vanish identically. From the equations above it follows that:

$$|f'(z)| \leqq 2\,\|f\|\,|z|,$$

$$\|f''(z)\| = 2\,\|f\|.$$

Proof: From the definition of β it follows that

$$f(z+h) - f(z) = f(h) + \beta(z, h)$$

and (4) implies that $f(h) = o(h)$. Since $\beta(z, h)$ is linear in h, it follows that f has an F-derivative at every z and the equality $f'(z)\,h = \beta(z, h)$ holds. Now $f' : B \to \mathrm{Hom}(B, V)$ (being equal to β) is obviously *linear*. Then from the general fact that the derivative of a linear mapping at any point is that linear mapping, we conclude that $f''(z) = f'$, or $f''(z) = \beta$.

Q.E.D.

CHAPTER II

Hard Implicit Functional Theorems

A. Newton's Method and the Nash Implicit Functional Theorem

The following theorem will be proved by the so-called Newton's method.

2.1. Theorem: Let B be a Banach space, and let f be a mapping whose domain $D(f)$ is the unit sphere of B. Suppose that:

(i) f has two continuous Fréchet derivatives in $D(f)$, both bounded above by a constant M, which we assume to exceed 2.

(ii) There exists a map $L(u)$ with domain $D(L) = D(f)$ and range in the space $\mathscr{B}(B)$ of bounded linear maps of B into itself, such that

(iia) $$|L(u)\,h| \leqq M\,|h|, \quad h \in B, \quad u \in D(L)$$

(iib) $$df(u)\,L(u)\,h = h, \qquad h \in B, \quad u \in D(L).$$

Then, if $|f(0)| < M^{-5}$, it follows that $f(D(f))$ contains the origin.

Proof: Let $\varkappa = \frac{3}{2}$, and let $\beta > 0$ be a real number to be specified later. Put $u_0 = 0$ and, proceeding inductively, put

(2.1) $$u_{n+1} = u_n - L(u_n)\,f(u_n).$$

We will prove inductively that

(2.2; n) $$u_{n-1} \in D(L)$$

(2.3; n) $$|u_n - u_{n-1}| \leqq e^{-\beta\varkappa^n}, \qquad n \geqq 1.$$

We proceed as follows. Suppose that statements (2.2; j) and (2.3; j) are true for $j \leqq n$. Then

$$|u_n| \leqq \sum_{j=1}^{\infty} e^{-\beta \varkappa j} \leqq \sum_{j=1}^{\infty} e^{-\beta(\varkappa-1)j} \leqq \frac{e^{-\beta(\varkappa-1)}}{1-e^{-\beta(\varkappa-1)}} \tag{2.4}$$

so that if β is sufficiently large, (2.2; n) follows. Therefore Definition (2.1) makes sense. Observe now that if g is any function twice continuously F-differentiable, the mean-value theorem with Lagrange remainder applied to $g(u + th)$ yields

$$g(u+h) = g(u) + dg(u)h + \int_0^1 (1-t)\, d^2 g(u+th, h, h)\, dt.$$

Combining this with (ii) and our induction hypothesis yields

$$\begin{aligned} |u_{n+1} - u_n| &= |L(u_n) f(u_n)| \leqq M\,|f(u_n)| \\ &\leqq M\,|f(u_{n-1}) - df(u_{n-1})\,L(u_{n-1}) f(u_{n-1})| \\ &\quad + M^2\,|u_n - u_{n-1}|^2 = M\,|u_n - u_{n-1}|^2 \leqq M^2 e^{-2\beta\varkappa^n}. \end{aligned} \tag{2.5}$$

Thus we have only to choose β so that

$$M^2 e^{-2\beta\varkappa^n} \leqq e^{-\beta\varkappa^{n+1}},$$

or

$$M^2 \leqq e^{(2-\varkappa)\beta\varkappa^n}. \tag{2.6}$$

Since $\varkappa < 2$, it is clear that (2.6) will hold for β sufficiently large, and then (2.3; n) follows, completing our induction. Thus we have only to prove the correctness of (2.3; 1) to finish our proof. But this statement is simply

$$|L(0) f(0)| \leqq e^{-\beta\varkappa} \tag{2.7}$$

and it is therefore implied by $M\,|f(0)| \leqq e^{-\beta\varkappa}$. Since $|f(0)| \leqq M^{-5}$, if we choose β so that $M^2 = e^{(1/2)\beta\varkappa}$, (2.7) follows. Then u_n converges to some element u in $D(f)$. By (iib) and (2.1)

$$f(u_n) = df(u_n)\,(u_n - u_{n+1}),$$

$$|f(u_n)| \leqq M\,|u_{n+1} - u_n| \leqq M\, e^{-\beta\varkappa^n},$$

so $f(u) = 0$.

Q.E.D.

The following theorem, which gives an important generalization of Theorem 2.1, is proved by a modified Newton's method. We weaken the hypotheses of Theorem 2.1 requiring not that the "inverting" operator $L(u)$ be bounded, but only that it be an unbounded operator acting somewhat like a differential operator of order α.

Given a compact n-dimensional manifold K we introduce the space $C^r(K) = C^r$ of (possibly vector-valued), r times continuously differentiable functions with the norm

$$|u|_r = \max_{|\alpha| \leqq r} \max_{x \in M} |D^\alpha u(x)|$$

$\alpha = (\alpha_1, \alpha_2, \ldots, \alpha_n)$, α_i non-negative integers, $|\alpha| = \alpha_1 + \cdots + \alpha_n$

$$D^\alpha = \left(\frac{\partial}{\partial x_1}\right)^{\alpha_1} \cdots \left(\frac{\partial}{\partial x_n}\right)^{\alpha_n}.$$

Note that $C^r \supseteq C^{r+1}$ and that, if $u \in C^{r+1}$, $|u|_r \leqq |u|_{r+1}$; we write $|u|_r = \infty$ if $u \notin C^r$. In the sequel we shall refer to a certain range $m - \alpha \leqq r \leqq m + 10\alpha$ of spaces C^r and to a certain constant $M \geqq 1$. We suppose that M is sufficiently large so that there exist *smoothing operators* $S(t)$, $t \geqq 1$ such that

(S$_1$) $$|S(t)\, u|_\varrho \leqq M t^{\varrho - r} |u|_r, \qquad u \in C^r$$

(S$_2$) $$|(I - S(t))\, u|_r \leqq M t^{r-\varrho} |u|_\varrho, \qquad u \in C^\varrho$$

(S$_3$) $$\left|\frac{d}{dt} S(t)\, u\right|_r \leqq M t^{r-\varrho-1} |u|_\varrho, \qquad u \in C^\varrho$$

(S$_4$) $$\lim_{t \to +\infty} |(I - S(t))\, u|_r = 0, \qquad i \in C^r$$

for $m - \alpha \leqq r \leqq \varrho \leqq m + 10\alpha$.

(We will show later how to construct these operators for any compact manifold.)

We proceed now to the statement of the main result of this chapter:

2.2. Nash implicit functional theorem: Let f be a mapping whose domain $D(f)$ is the unit sphere of C^m with range in $C^{m-\alpha}$. Suppose that

(i) f has two continuous F-derivatives, both bounded by M.

(ii) There exists a map $L(u)$ with domain $D(L) = D(f)$ and range in the space $\mathscr{B}(C^m, C^{m-\alpha})$ of bounded linear operators on C^m to $C^{m-\alpha}$, such that:

(iia) $|L(u)h|_{m-\alpha} \leqq M|h|_m$, $\qquad u \in D(L),\ h \in C^m$

(iib) $df(u)L(u)h = h$, $\qquad u \in D(L),\ h \in C^{m+\alpha}$

(iic) $|L(u)f(u)|_{m+9\alpha} \leqq M(1 + |u|_{m+10\alpha})$, $\qquad u \in C^{m+10\alpha}$.

Then, if

$$|f(0)|_{m+9\alpha} \leqq 2^{-40} M^{-202},$$

$f(D(f))$ contains the origin.

Proof: Let $\varkappa = \frac{3}{2}$ and $\beta, \mu, \nu > 0$ be real numbers to be specified later. Put $u_0 = 0$, and proceeding inductively, put

$$\text{(2.8; n)} \qquad u_{n+1} = u_n - S_n L(u_n) f(u_n)$$

where $S_n = S(e^{\beta\varkappa^n})$. We will prove inductively that

$$\text{(2.9; n)} \qquad u_n \in D(f)$$

$$\text{(2.10; n)} \qquad |u_n - u_{n-1}|_m \leqq e^{-\mu\alpha\beta\varkappa^n}$$

$$\text{(2.11; n)} \qquad u_n \in C^{m+10\alpha}$$

$$\text{(2.12; n)} \qquad 1 + |u_n|_{m+10\alpha} \leqq e^{\nu\alpha\beta\varkappa^n}.$$

Suppose (2.10; j) is true for $j \leqq n$. Then

$$|u_n|_m \leqq \sum_{j=1}^{\infty} e^{-\mu\alpha\beta\varkappa^j} \leqq \sum_{j=1}^{\infty} e^{-\mu\alpha\beta(\varkappa-1)j} = \frac{e^{-\mu\alpha\beta(\varkappa-1)}}{1 - e^{-\mu\alpha\beta(\varkappa-1)}}$$

which implies (2.9; n) if β, μ are sufficiently large. Suppose now that (2.9; j), (2.10; j), (2.11; j), (2.12; j) are true for $j \leqq n$. Then

$$\begin{aligned}
|u_{n+1} - u_n|_m &= |S_n L(u_n) f(u_n)|_m \\
&\leqq M e^{\alpha\beta\varkappa^n} |L(u_n) f(u_n)|_{m-\alpha} \leqq M^2 e^{\alpha\beta\varkappa^n} |f(u_n)|_m \\
&\leqq M^2 e^{\alpha\beta\varkappa^n} |f(u_{n-1}) - df(u_{n-1}) S_{n-1} L(u_{n-1}) f(u_{n-1})|_m \\
&\quad + M^3 e^{\alpha\beta\varkappa^n} |u_n - u_{n-1}|_m^2 \\
&\leqq M^2 e^{\alpha\beta\varkappa^n} |df(u_{n-1}) (I - S_{n-1}) L(u_{n-1}) f(u_{n-1})|_m \\
&\quad + M^3 e^{\alpha\beta\varkappa^n} e^{-2\mu\alpha\beta\varkappa^n} \\
&\leqq M^3 e^{\alpha\beta\varkappa^n} [M e^{-9\alpha\beta\varkappa^{n-1}} |L(u_{n-1}) f(u_{n-1})|_{m+9\alpha} + e^{-2\mu\alpha\varkappa^n}] \\
&\leqq M^4 e^{\alpha\beta\varkappa^n} [e^{-9\alpha\beta\varkappa^{n-1}} M(1 + |u_{n-1}|_{m+10\alpha}) + e^{-2\mu\alpha\beta\varkappa^n}] \\
&\leqq M^5 e^{\alpha\beta\varkappa^n} [e^{-9\alpha\beta\varkappa^{n-1}} e^{\nu\alpha\beta\varkappa^{n-1}} + e^{-2\mu\alpha\beta\varkappa^n}] \\
&\leqq M^5 \{\exp \alpha\beta\varkappa^{n-1} (\nu - 9 + \varkappa) + \exp \alpha\beta\varkappa^n (1 - 2\mu)\}.
\end{aligned}$$

The desired inequality will be then implied by

$$(2.13; \mathrm{n}) \qquad M^5 \{\exp [\alpha\beta\varkappa^{n-1} (\nu - 9 + \varkappa)] + \exp [\alpha\beta\varkappa^n (1 - 2\mu)]\} \leqq e^{-\mu\alpha\beta\varkappa^{n+1}}$$

which (noting that $\varkappa = \frac{3}{2}$) will follow for β sufficiently large if we choose

$$(2.14) \qquad \mu > 2, \quad \tfrac{9}{4}\mu + \nu < \tfrac{15}{2}.$$

Thus (2.10; n + 1) follows.

Next we note that

$$(2.15) \qquad \begin{aligned} 1 + |u_{n+1}|_{m+10\alpha} &\leqq 1 + \sum_{j=0}^{n} |S_j L(u_j) f(u_j)|_{m+10\alpha} \\ &\leqq 1 + M \sum_{j=0}^{n} e^{\alpha\beta\varkappa^j} |L(u_j) f(u_j)|_{m+9\alpha} \\ &\leqq 1 + M^2 \sum_{j=0}^{n} e^{\alpha\beta\varkappa^j} (1 + |u_j|_{m+10\alpha}) \\ &\leqq 1 + M^2 \sum_{j=0}^{n} e^{\alpha\beta(1+\nu)\varkappa^j}. \end{aligned}$$

Thus

$$(2.16) \qquad \begin{aligned} &(1 + |u_{n+1}|_{m+10\alpha})\, e^{-\nu\alpha\beta\varkappa^{n+1}} \\ &\leqq e^{-\nu\alpha\beta\varkappa^{n+1}} + M^2 \sum_{j=0}^{n} e^{\alpha\beta\varkappa^j(1+\nu-\varkappa\nu)}. \end{aligned}$$

If $\nu > 2$ the right side of (2.16) will be less than 1 for sufficiently large β, and so statement (2.12; n + 1) will follow from (2.16), completing our induction. If we take $\nu = \frac{5}{2}$, $\mu = \frac{11}{5}$, condition (2.14) is satisfied and so we have only to verify the correctness of statements (2.10; 1) and (2.12; 1) and our proof will be complete. These statements, however, are simply the inequalities

$$(2.17) \qquad |S_1 L(0) f(0)|_m \leqq e^{-\mu\alpha\beta\varkappa}$$

and

$$(2.18) \qquad 1 + |S_1 L(0) f(0)|_{m+10\alpha} \leqq e^{\nu\alpha\beta\varkappa}$$

and they in turn follow from the bound for $|f(0)|$ and (iic). The conclusion of the proof is now just as in Theorem 2.1.

We proceed now to the construction of the family of smoothing operators whose existence was assumed for the special case $K = n$-dimensional torus;

then $C^k(K)$ is simply the space of all k-times differentiable functions $u(x)$, defined in E^n and periodic with period 2π in each variable. Take a sufficiently large constant M and a function $\hat{a} \in C^\infty(E^m)$, vanishing outside a compact set and identically equal to 1 in a neighborhood of 0, and let a be its Fourier transform. It is well known that for any α, N

$$|D^\alpha a(x)| \leqq A_{\alpha,N}(1 + |x|)^{-N}.$$

Moreover

$$\int_{E^n} a(x)\, dx = 1,$$

$$\int_{E^n} x^\alpha a(x)\, dx = 0, \quad x^\alpha = x^{\alpha_1} x^{\alpha_2} \cdots x^{\alpha_n}, \quad |\alpha| > 0.$$

Now we set

$$(S(t)\, u)(x) = t^n \int_{E^n} a(t(x - y))\, u(y)\, dy.$$

It is clear that $S(t)\, u \in C^\infty$ and, since $S(t)$ commutes with partial differentiation operators, we have to prove statements (S_1), (S_2) above only for $r = 0$. In fact, suppose (S_1) is true for $r = 0$. Then

$$|S(t)\, u|_{\varrho - r} \leqq M t^{\varrho - r} |u|_0 .$$

Taking any α, $|\alpha| \leqq r$

$$|D^\alpha S(t)\, u|_{\varrho - r} = |S(t)\, D^\alpha u|_{\varrho - r}$$
$$\leqq M t^{\varrho - r} |D^\alpha u|_0 \leqq M t^{\varrho - r} |u|_r .$$

But then

$$|S(t)\, u|_r \leqq M t^{\varrho - r} |u|_r .$$

One deals similarly with (S_2).

Suppose then that $r = 0$. (S_1) reduces to

$$|S(t)\, u|_\varrho \leqq M t^\varrho |u|_0 .$$

Let $|\alpha| \leqq \varrho$. We have

$$|D^\alpha S(t)\, u|_0 = \left| t^{n+|\alpha|} \int_{E^n} D^\alpha a(t(x - y))\, u(y)\, dy \right|$$
$$\leqq M t^{|\alpha|} |u|_0 \leqq M t^\varrho |u|_0$$

if $|\alpha| \leqq \varrho$, so (S_1) is established. (S_2) reduces to

$$|(1 - S(t))\, u|_0 \leqq M t^{-\varrho} |u|_\varrho .$$

To prove this, apply Taylor's theorem with integral remainder:

$$f(1) = \sum_{k=0}^{m-1} \frac{f^{(k)}(0)}{k!} + \frac{1}{(m-1)!}\int_0^1 (1-\mu)^{m-1} f^{(m)}(\mu)\, d\mu$$

to the function $f(t) = u(x + ty)$. We obtain

$$u(x+y) = \sum_{k=0}^{\varrho-1} \frac{1}{k!}\Big(\sum_{|\alpha|=k} y^\alpha D^\alpha u(x)\Big)$$

$$+ \frac{1}{(\varrho-1)!}\sum_{|\alpha|=\varrho} y^\alpha \int_0^1 (1-\mu)^{\varrho-1} D^\alpha u(x+\mu y)\, d\mu.$$

Thus

$$u - S(t)u = t^n \int_{E^n} a(t(x-y))(u(x) - u(y))\, dy$$

$$= \frac{t^n}{(\varrho-1)!}\sum_{|\alpha|=\varrho}\int_{E^n}\int_0^1 a(t(x-y))(1-\mu)^{\varrho-1} y^\alpha D^\alpha u(x+\mu y)\, d\mu\, dy.$$

Making the change of variable $ty = z$, we obtain

$$t^n \int_{E^n}\int_0^1 a(tx - ty)(1-\mu)^{\varrho-1} y^\alpha D^\alpha u(x+\mu y)\, d\mu\, dy$$

$$= t^{-|\alpha|}\int_{E^n}\int_0^1 a(tx - z)(1-\mu)^{\varrho-1} z^\alpha D^\alpha u(x+\mu t^{-1}z)\, d\mu\, dz.$$

But then it is easy to conclude

$$|I - S(t)u|_0 \leqq Mt^{-\varrho}|u|_\varrho$$

which proves (S_2).

Let us pass now to (S_3). As before, we can suppose without loss of generality that $r = 0$, in which case, (S_3) reduces to

$$\left|\frac{d}{dt}S(t)u\right|_0 \leqq Mt^{-\varrho-1}|u|_0.$$

But

$$\frac{d}{dt}S(t)u = \frac{d}{dt}t^n\int_{E^n} a(t(x-y))u(y)\, dy$$

$$= t^{n-1}\int_{E^n}\Big(na(t(x-y)) + \sum_{i=1}^n ty^i (D^i a)(t(x-y))\Big)u(y)\, dy.$$

Reasoning entirely analogous to that used in the proof of (S_2) yields the desired result. As for (S_4), it is a well-known result in the theory of singular integrals, and therefore we omit the proof.

We note that the construction of the smoothing operators could be carried out for any compact manifold, and not only for the torus. For the proof, we refer the reader to J. Schwartz, *On Nash's Implicit Functional Theorem*, Comm. Pure Appl. Math., vol. 13 (1960), pp. 509–530. We note also that the use of spaces C^r and the norms $|\cdot|_r$ is by no means essential in the proof of Nash's implicit functional theorem; indeed, these spaces can be replaced, for example, by spaces like $L_p^r(K) = L_p^r$ = space of all (possibly vector-valued) functions f for which

$$\int_K |D^\alpha f|^p \, dx < \infty,$$

$|\alpha| \leqq r$, with the norm

$$|f|_r^p = \sum_{|\alpha| \leqq r} \int_K |D^\alpha f|^p \, dx.$$

We present now a useful corollary of Nash's Implicit Functional Theorem.

2.3. 2nd implicit functional theorem: Let T = n-dimensional torus, let $f\colon C^k \to C^{k-\beta}$ be defined on the unit sphere of C^k, and suppose that

(i) f has infinitely many continuous F-derivatives.

(ii) f is *translation invariant*, i.e. if $u \in C^k$, $|u|_k < 1$

$$f(u(\cdot + h))(s) = \{f(u)\}(x + h).$$

(iii) There exists a mapping $L(u)$ defined in the unit sphere of C^k with values in $\mathscr{B}(C^k, C^{k-\beta})$ such that $L(u)$ is *translation invariant* in the same sense as f, such that $L(u)$ has infinitely many continuous F-derivatives, and such that

(iiia) $$|L(u)\,h|_{k-\beta} \leqq M\,|h|_k, \qquad u \in C^k, \qquad h \in C^k$$

(iiib) $$df(u)\,L(u)\,h = h, \qquad u, h \in C^k.$$

Then, if $f(0) = 0$, $f(D(f))$ contains a C^∞-neighborhood of zero.

Proof: Note that, since f is translation invariant, it commutes with derivatives, so if we apply f to a function in $C^{k'}$, $k' > k$, we obtain a function in $C^{k'-\beta}$; similarly $L(u)$ can be considered as a function whose domain is the unit sphere of $C^{k'}$ and range $C^{k'-\beta}$. The inequalities and identities

(iiia)* $$|L(u)\,h|_{k'-\alpha} \leqq M\,|h|_{k'}, \qquad u \in C^{k'}, \qquad h \in C^{k'}$$

(iiib)* $$df(u)\,L(u)\,h = h, \qquad u, h \in C^{k'+\beta}$$

also hold. We now have only to apply Nash's implicit functional theorem, for which we need (iic) of its statement. But this is a consequence of the translation-invariance of f and L together with inequality (iiia) and the boundedness of the derivatives of f. Applying Nash's implicit functional theorem, it follows that if a point k is sufficiently near to the origin in $C^{k-\beta}$, there is a point in C^k whose image is k. Therefore, $f(D(f))$ contains a $C^{k-\beta}$-neighborhood of the origin, and thus a C^∞-neighborhood also.

We show now how the implicit functional theorem can be applied, first to an artificial example and then to a natural one.

B. A Partial Differential Equation

Consider functions of n variables, of period 2π in each variable, i.e. functions on the n-dimensional torus. The partial differential operator

$$\square = \left(\frac{\partial}{\partial x_1}\right)^4 - \left(\frac{\partial}{\partial x_2}\right)^4 + i\left(\frac{\partial}{\partial x_3}\right)^4 + \left(\frac{\partial}{\partial x_4}\right)^2 - \left(\frac{\partial}{\partial x_5}\right)^2 - \cdots - \left(\frac{\partial}{\partial x_n}\right)^2$$

has, by deliberate choice, an extremely unfortunate "mixed" character from the point of view of the theory of partial differential operators. But it is easy to see that $\square$ admits the complete orthogonal set of functions

$$\exp\left(i\left(m_1x_1 + \cdots + m_nx_n\right)\right)$$

as eigenfunctions, and that the eigenvalues of $\square$ are Gaussian integers. Therefore, the equation

$$\text{(B1)} \qquad (\square + \tfrac{1}{2})\, u = v$$

is invertible in the following sense: if v is a function in $L^2(K)$, then there is a function $u \in L^2(K)$ such that (B1) is valid in the L^2-sense.

Our aim is to show that the equation

$$f(u) = \square u + \tfrac{1}{2}u + u^3 \exp(\square u) = v$$

has a solution $u \in C^\infty$ for sufficiently small v in C^∞. Observe first that f maps C^k into C^{k-4} for any k and that it has infinitely many continuous F-derivatives, the first of which is

$$df(u)\,h = (1 + u^3 \exp(\square u))\,\square h + (\tfrac{1}{2} + 3u^2 \exp(\square u))\,h.$$

To find $L(u)$ we have to invert

$$\text{(B2)} \qquad \Box h + \frac{\frac{1}{2} + 3u^2 \exp(\Box u)}{1 + u^3 \exp(\Box)} h = \frac{l}{1 + u^3 \exp(\Box u)}.$$

For $u = 0$, (B2) reduces to (B1), so by a standard perturbation theory argument, (B2) will be invertible for any u sufficiently close to zero in C^k. The operator $L(u)\, l = h$ will then be defined and certainly continuous as an operator from C^k to C^{k-4}; thus inequality (iiia) follows for L, and (iiib) is an immediate consequence of the definition of u, i.e., of the fact that $L(u)$ has infinitely many derivatives and is translation invariant. But we have now verified all the hypotheses of Theorem 2.3, so our result follows at once.

We note next that our "translation invariance" requirements on f do not prevent us from treating some apparently unmanageable cases, such as

$$f(u) = \Box u(x) + c_1(x)\, u(x) + c_2(x)\, u^3(x) \exp(\Box u(x)) = v(x),$$

where $c_1(x)$ and $c_2(x)$, are C^∞ functions on the n-dimensional torus. In fact, we have only to look at this problem as if it were that of solving the system of equations

$$\Box u + d_1 u + d_2 u^3 \exp(\Box u) = v$$

$$d_1 = c_1$$

$$d_2 = c_2 .$$

If we suppose that the operator $u \to (\Box + c_1)\, u$ has a bounded inverse in L_2 (as was the case for $c_1 = \frac{1}{2}$), then the first Frechet derivative of the infinitely differentiable mapping

$$F\colon [d_1, d_2, u] \to [\Box u + d_1 u + d_2 u^3 \exp(\Box u), d_1, d_2]$$

will have an inverse; in fact

$$dF[d_1, d_2, u][s_1, s_2, h] = [\Box h + s_1 u + d_1 h + s_2 u^3 \exp(\Box u)$$
$$+ 3 d_2 u^2 \exp(\Box h) + d_2 u^3 \exp(\Box u)\, \Box h, s_1, s_2]$$

which, for d_1 and d_2 near c_1 and c_2 and f sufficiently close to zero in suitable senses may be solved as in the previous case. Observing that F is translation-invariant, and reasoning as before, our result follows.

C. Embedding of Riemannian Manifolds

Bibliography

1. N. Bourbaki, *Espaces vectoriels topologiques.*
2. S. Helgason, *Differential Geometry and Symmetric Spaces.*
3. S. Lang, "Fonctions Implicites et plongements Riemanniens", Sém. Bourbaki, E.N.S., exposé **237** (1961–62).
4. J. Nash, "The imbedding problem for Riemannian manifolds", Ann. of Math. vol. 63, pp. 20–63 (1956).

Now we shall consider the problem of isometric embeddings of Riemannian manifolds in euclidean spaces. This problem was successfully treated for the first time by John Nash (see [4]), and it provides a natural application for theorems such as Theorem 2.2 above. The problem can be stated as follows. Is every Riemannian manifold (say of class C^k) isometrically embeddable in R^m? (Throughout this section, *embedding* means *diffeomorphic mapping with injective differential at each point* (= *regular* at each point).) Nash's answer is in the affirmative (technically, when $k \geqq 3$), and he also asserts that m may be chosen less than or equal to an explicit function of the dimension n of the manifold (namely $m \leqq \frac{1}{2}(3n^3 + 14n^2 + 11n)$ for the general case and $m \leqq \frac{1}{2}n(3n + 11)$ if M is compact). Actually we shall here prove only a weak result: our final statement will deal only with C^∞-compact Riemannian manifolds and no bounds for m will be determined. M will henceforth denote a *compact Riemannian manifold* of dimensoin n. The manifold itself and its metric will be supposed of class C^∞.

I. Remark: Without loss of generality the manifold M may be supposed to be a torus ([3], No. 1). In fact, by Whitney's theorem (cf. G. de Rham, *Variétés differéntiables*, or Milnor, *Notes on Differential Topology*, Princeton, 1959) M can be represented as a closed smooth bounded submanifold of some Euclidean space E^N. But then by properly choosing everything we can assume that the projection of E^N on some torus is $1 - 1$ on the manifold M. This represents M as a closed smooth submanifold of a torus.

Now we have some Riemannian metric defined on the submanifold M of the torus. By a standard procedure using partitions of unity it is possible to extend this metric to a metric on all of the torus. If we now isometrically embed the torus equipped with this metric we obtain by restriction an isometric embedding of M. Thus we may always suppose that our manifold is a torus; this will simplify some constructions. Nevertheless we begin by discussing an arbitrary compact manifold, because as far as VI below the assumption that M is a torus makes no difference in the proofs.

II. Consider the Banach space $C^r(M, R^m)$ of r-times differentiable functions on M with values in R^m, and, more generally the Banach space S^r of symmetric, doubly covariant, r times continuously differentiable tensor fields on M, defined by dealing locally with matrices instead of with real numbers (see above). Such tensor fields are metrics and R^m has a canonical metric, namely the Euclidean metric. Each $z \in C^r(M, R^m)$ induces "by devolution" of this metric an element of S^{r-1}, and therefore we have a mapping f: $C^r(M, R^m) \to S^{r-1}$ (for a more explicit definition see below). We shall show that for m large enough, the image of f covers an open set of S^∞. To do so we shall prove the hypothesis of Theorem 2.2, and then establish our claim easily.

III. First of all we want to know the Frechet derivatives of f (and f itself). Suppose that $z_1, \ldots, z_m$ are the canonical coordinates in R^m, and that $x_1, \ldots, x_n$ is a coordinate system defined on some open set U of D. Then if $z \in C^r(M \subset R^m)$ and $f(z)$ denotes, as above, the tensor on M induced by z, we have the following expression in coordinates:

$$(1) \qquad (f(z))_{i,j} = \sum_\alpha \frac{\partial z_\alpha}{\partial x_i} \frac{\partial z_\alpha}{\partial x_j}.$$

This formula is standard and may be taken as the starting point, but (at the risk of being more boring than necessary) we add the following exposition.
If X is a manifold and $p \in X$, denote by TX_p the tangent space to X at p. Now if $z : M \to R^m$ is smooth, $p \in M$, $q = z(p)$, z has a differential at p, i.e., z induces a map

$$z^* : TM_p \to T(R^m)_q$$

which is linear (see [2], § 3, No. 1). But the metric on R^m induces an isomorphism

$$u : T(R^m)_q \to (T(R^m)_q)^* \qquad (* = \text{dual space}).$$

Consider now the linear mapping A obtained by composition of the mappings:

$$TM_p \xrightarrow{z^*} T(R^m)_q \xrightarrow{u} (T(R^m)_q)^* \xrightarrow{{}^tz^*} (TM_p)^*,$$

where ${}^tz^*$ stand for the transpose of z^*. Clearly $A : TM_p \to (TM_p)^*$. As there exists a canonical identification:

$$Hom_R(TM_p, (TM_p)^*) = (TM_p)^* \underset{R}{\otimes} (TM_p)^*,$$

A may be considered as a doubly covariant tensor at p. The correspondence $z \to A$ is what we called f, i.e. we *define* f by $(f(z))_p = A$. Observe that the fact that the values of z are in R^m has not played any special role, and any Riemannian manifold could replace R^m. But in our case, we know that $T(R^m)_p$ and $(T(R_m)_p)^*$ may be identified with R^m itself, and that the isomorphism u is the identity. Finally we get

(1a) $$(f(z))_p = z_p^* \cdot {}^t z_p^* .$$

In terms of coordinates, z^* is the Jacobian matrix $z^* = J_z = \left(\frac{\partial z_i}{\partial x_j}\right)$, and from (1a) it follows that

(1b) $$f(z) = J_z \cdot {}^t J_z .$$

But then

$$(f(z))_{i,j} = \sum_{\alpha} \frac{\partial z_\alpha}{\partial x_i} \frac{\partial z_\alpha}{\partial x_j},$$

which is (1) above. From formula (1a) or (1b) it follows at once that $f\colon C^{r+1} \to S^r$ is a *quadratic form* (see 2, Chap. I); the bilinear form β associated with f is

(2) $$\beta(x, y) = x^* \cdot {}^t y^* + y^* \cdot {}^t x^* .$$

Another consequence of (1b) is the continuity of f (as a function from C^{r+1} into S^r). This is clear. We may therefore apply Theorem 1.50 and conclude that f has derivatives of all orders:

(3) $$f'(z)\, h = z^* \cdot {}^t h^* + h^* \cdot {}^t z^*,$$

$$f''(z)(h, k) = h^* \cdot {}^t k^* + k^* \cdot {}^t h^*,$$

$$f^{(n)}(z) = 0 \quad \text{if} \quad n \geqq 3,$$

and that the norms satisfy

(3′) $$|f'(z)| \leqq 2\,\|f\|\,|z|$$

$$\|f''(z)\| = 2\,\|f\| .$$

In terms of coordinates, (3) may be written as:

(3a) $$f'(z)\, h = J_z {}^t J_h + J_h {}^t J_z ,$$

(3b) $$(f(z)\, h)_{ij} = \sum_{\alpha} \frac{\partial z_\alpha}{\partial x_i} \frac{\partial h_\alpha}{\partial x_i} + \frac{\partial z_\alpha}{\partial x_j} \frac{\partial h_\alpha}{\partial x_i} .$$

Naturally we plan to show that $f'(z)\,h$ is invertible (as a function of h) in a very smooth way, i.e.,

(4) given $g \in S^r$ and $z \in C^s$ we want to find h as differentiable as possible such that $g = f'(z)\,h$.

IV. To achieve this we use a trick invented by Nash ([4], p. 31) which to the problem of solving (4) adds some new conditions. In other words, we require that the solution h have an additional property given a priori, namely

$$ {}^t z_p^* \,(h(p)) = 0 \quad \text{for every} \quad p \in M. \tag{5}$$

Since $(T(R^m)_p)^*$ was identified with R^m and

$$ {}^t z_p^* : (T(R^m)_q)^* \to (TM_p)^*, \qquad h(p) \in R^m, $$

it is clear that (5) makes sense.

Of course (4) and (5) may be written in coordinates as:

$$ g_{ij} = \sum_\alpha \frac{\partial z_\alpha}{\partial x_i}\frac{\partial h_\alpha}{\partial x_j} + \frac{\partial z_\alpha}{\partial x_j}\frac{\partial h_\alpha}{\partial x_i}, \tag{4a}$$

$$ \sum_\alpha \frac{\partial z_\alpha}{\partial x_i} h_\alpha = 0, \quad i = 1, \ldots, n. \tag{5a}$$

We now prove that the conditions (4) and (5) may be satisfied simultaneously by a suitable h. From (5a) we conclude that

$$ \sum_\alpha \frac{\partial z_\alpha}{\partial x_i}\frac{\partial h_\alpha}{\partial x_j} + \frac{\partial^2 z}{\partial x_i \partial x_j} h_\alpha = 0 $$

or

$$ \sum_\alpha \frac{\partial z_\alpha}{\partial x_i}\frac{\partial h_\alpha}{\partial x_j} = -\sum_\alpha \frac{\partial^2 z_\alpha}{\partial x_i\, \partial x_j}, $$

and then (4a) becomes

$$ g_{ij} = -2\sum_\alpha \frac{\partial^2 z_\alpha}{\partial x_i\, \partial x_j} h_\alpha . \tag{4b}$$

This shows the point of adding the condition (5): now (4b) and (5) give a *system of algebraic linear equations*, equivalent to (4) and (5), which are a system of partial differential equations of first order.

Equations (4) and (5) (or (4b) and (5a)) can be written for every $z \in C^{r+1}(M, R^m)$. Nevertheless, we can assure the existence of a solution only for a non-void open set of z's in $C^3(M, R^m)$, where m is large enough (m may be chosen to be $m = 2n^2 + 3n$ – see [4], p. 53—but remember that we don't care about bounds for m).

V. Choose a mapping of M into R^s by functions $v_1, \ldots, v_s$. Now define a mapping $\bar{z}$ of M into $R^{s+(1/2)s(s+1)}$ by means of the functions

$$\begin{array}{l} v_1 \quad , \ldots, v_s \\ v_1^2 \quad , \ldots, v_1 v_s \\ v_2 v_1, \ldots, v_2 v_s \\ \vdots \\ v_s v_1, \ldots, v_s^2. \end{array}$$

Write

$$\bar{z}_1 = v_1, \bar{z}_2 = v_2, \ldots, \bar{z}_s = v_s, \quad \bar{z}_{i,j} = v_i v_j, \quad i \leqq j.$$

If $v = (v_1, \ldots, v_s)$ is a *regular* C^∞ *embedding* of M into R^s (the existence of such is guaranteed by Whitney's Theorem), we claim that $f'(z)\,h$ is invertible as a function of h for every z in a neighborhood of $\bar{z}$. In fact, if $v = (v_1, \ldots, v_s)$ is a regular embedding, one can take as local coordinates for M some appropriate subset of the v_i's. Suppose that M has been covered by the open sets $U_1, \ldots, U_N$ in such a way that on each U_i one such subset works as a coordinate system. We consider the linear systems (4b) and (5a) in this particular case. In order to simplify the notation, order the z's once and for all by

$$(z_1, \ldots, z_m, z_{1,1}, z_{1,2}, \ldots, z_{s,s})$$

and write α as a general index for them. Let us fix one particular U_i (call it simply U) and suppose that $x_1 = v_1, x_2 = v_2, \ldots, x_n = v_n$ is a system of coordinates on U. Consider $\bar{z}$. The coefficients of (4b) and (5a) are first or second derivatives of the $\bar{z}_\alpha$'s with respect to the x_i's, and the matrix B of the system of linear equations has the following form at every point of U:

B=
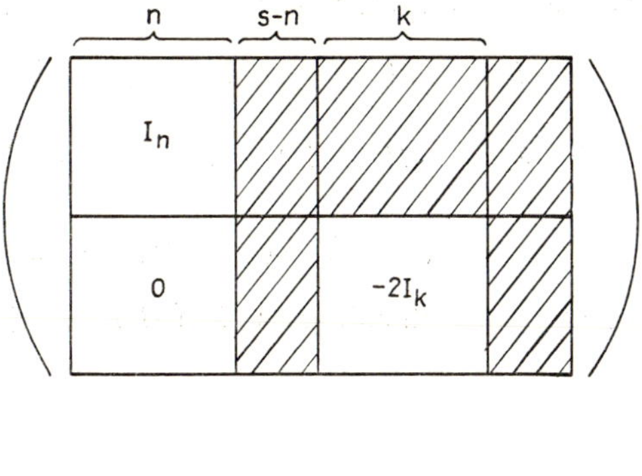

In the above, shading indicates arbitrary coefficients and $k = \frac{1}{2}n(n+1)$. I_n, I_k are the identity matrices of dimension and k respectively.

This shows that the matrix B has maximal rank $n + \frac{1}{2}n(n+1)$ at every point of U (its rows being linearly independent). The same is true (for obvious reasons) for every z which is sufficiently near $\bar{z}$ in the C^2 sense.

This remark has two strong consequences. First, it clearly shows that the systems (4b) and (5a) have solutions at every point of U for all z, C^2 near $\bar{z}$. This is basic in finding h. Second, it will follow from this that there exists a solution of (4b) and (5a) defined by a mapping that is as smooth as g and the second derivatives of z are. This needs some explanation. At each point of U, we know that (4b) and (5a) can be solved. Among the solutions of this system of linear equations we pick out one by the condition

(6)
$$\sum_{\alpha} (h_{\alpha})^2 = \text{minimum}.$$

It is easy to conclude from the fact that the solutions form a convex set in Euclidean space, that one well defined solution is thereby selected.

This defines a mapping $h: U \to R^l$, $l = s + \frac{1}{2}s(s+1)$, and two problems arise. (a) Is h differentiable? (b) Is h defined independently of U?

(a) Differentiability of h. Fix a point p in U. Since B has maximal rank, $B \cdot {}^tB$ is non-singular (this follows from the fact that $\det(B \cdot {}^tB)$ is the Gram determinant of the rows of B, and consequently different from zero if these rows are linearly independent). But then the equation

$$(B \cdot {}^tB)\, D = G$$

has a unique solution $D = (B \cdot {}^tB)^{-1} \cdot G$, for all $G = (g_{ij})$.

Define

(7)
$$H = {}^tBD = {}^tB\,(B \cdot {}^tB)^{-1} \cdot G.$$

Clearly H is a solution of (4b) and (5a) at the given point of U. We claim that this is the solution satisfying (6). In fact, if $\bar{H}$ is another solution, $B\bar{H} = G$, we have (writing (,) for the scalar product in R^l):

$$(\bar{H}, \bar{H}) - (H, H) = (\bar{H} - H, \bar{H} - H) + 2\,(H, \bar{H} - H)$$

and

$$(H, \bar{H} - H) = ({}^tBD, \bar{H} - H) = (D, B\bar{H} - BH) = (D, G - G) = 0.$$

Then

$$(\bar{H}, \bar{H}) - (H, H) = (\bar{H} - H, \bar{H} - H),$$

and this proves that among all solutions $\bar{H}$ of $B\bar{H} = G$, $(\bar{H},\bar{H})$ is minimum when $\bar{H} = H$.

If the point p in U is now permitted to vary, formula (7) shows that H varies as smoothly as B and G do. In it follows that h is r times differentiable provided that z is $t + 2$ times differentiable and g is r times differentiable.

(b) It remains to show that the solution H is independent of the coordinates chosen in U. But that is easy. If h_1, h_2 are solutions constructed on U_1 and U_2 respectively, since property (6) is coordinate independent, h_1 and h_2 both must possess it at every point of $U_1 \cap U_2$; by uniqueness h_1 and h_2 agree there. This proves that h may be defined everywhere, and thus we are through.

The expression

$$H = {}^tB\,(B \cdot {}^tB)^{-1}\,G \tag{7}$$

for H at p in terms of G at p (and z locally), assures that the correspondence $g \to h$ (z supposed fixed) is *linear*.

But it tells us still more. In fact, the matrix B involves first and second derivatives of z (B is the matrix of (4b) and (5a)). Then from (7) it is apparent that H is a smooth function of g and of the first and second derivatives of z.

Let $L(z)\,g$ denote the function H of (7). The smoothness of H as a function of z and g may be stated as follows: *the mapping* $L(z)\,g$ *is continuous in both variables* z *and* g *simultaneously for the topologies:* $z \in C^{r+2}(M, R^m)$, $g \in S^r$ $L(z)\,g \in C^r(M, R^m)$.

Naturally the equivalence between (4) and (5) and (4b) and (5) can be proved so long as h at least first derivatives (otherwise (4) does not make sense). For that we need $L(z)\,g$ to belong at least to $C^1(M, R^m)$. This requires r to be 3 or more.

We sum up as follows:

(8) *For every* $r \geqq 3$ *there exists an open set* $\mathcal{O}$ *in* $C^{r+2}(M, R^m)$ *such that a function* $L(z)\,g$ *is defined on* $\mathcal{O} \times S^r$ *is continuous (in both variables simultaneously), and has values in* $C^r(M, R^m)$ *satisfying:*

(8i) $f'(z) \circ L(z)\,g = g, \qquad (z, g) \in \mathcal{O} \times S^r$;

(8ii) $L(z)\,g$ *is linear in* g *for every fixed* z *in* $\mathcal{O}$;

(8iii) *the elements in* $\mathcal{O}$ *are* $1 - 1$ *and regular at every point.*

(8iii) follows from the fact that we may choose $\mathcal{O}$ to be a small neighborhood of the embedding $\bar{z}$ above, and that the set of embeddings is open in C^r, $r \geqq 1$.

VI. We now assume that M is a torus (cf. I). Hence there exists a global set of (local) coordinates (the angular parameters $x_1, \ldots, x_n$) and consequently there is a standard way of expressing the doubly covariant tensor fields as $n \times n$ matrices of functions, just by taking the components of such tensors in coordinates $x_1, \ldots, x_n$ at each point. This means that it is possible to identify S^r with $(C^r(M, R))^{n^2}$. But then the mapping $f: C^{r-1}(M, R^m) \to S^r$ may be assumed to have range in $(C^r(M, R))^{n^2}$ and hence to split in n^2 components each with range in $C^r(M, R)$.

Thus each component inherits from f all properties vis-à-vis derivatives. We leave to the reader now the verification that these components satisfy the hypotheses of Theorem 2.2.

We can then apply Theorem 2.2 and conclude that:

(10) *The image under f of the set of infinitely often continuously differentiable embeddings of M in some Euclidean space covers an open set of S^∞.*

Remark: We may restrict our embeddings to be embeddings of M in some fixed R^m, and the conclusion should remain the same. But our next step will consist of adding directly two such embeddings and the bound m will vanish. For that reason (9) is stated without any reference to the ranges of the embeddings considered.

VII. Let K^r = the set of all tensor fields in S^r that are metrics, that is to say positive definite at each point of M. Clearly K^r is a convex cone, open for the S^r topology.

For every C^{r+1} embedding z of M in some Euclidean space, $f(z)$ belongs to K^r. Let $E^r \subset K^r$ be the set of all such $f(z)$.

Lemma: E^∞ is a convex cone dense in K^∞ for the S^∞ topology.

Proof: (i) E^r is a convex cone. For every $\lambda \geqq 0$ we have $\lambda f(z) = f(\sqrt{\lambda} z)$ (f is a quadratic form). If $z: M \to R^m$, $u: M \to R_r$, then the embedding $t = z \oplus u: M \to R^m \oplus R^r$ (defined in the obvious way) satisfies $f(t) = f(z) + f(u)$. Both properties together define a convex cone.

(ii) E^∞ is dense in K^∞.

Proof: Suppose the contrary, and let $\bar{E}^\infty$ be the closure of E^∞. Then there exists a point $g \in K^\infty$ such that $g \notin \bar{E}^\infty$. By the separating hyperplane property of locally convex F-space S^∞ of all C^∞ tensors on the manifold M, there exists a continuous linear functional ϕ on S^∞ such that $\phi(\bar{E}^\infty) \leqq 0$ and $\phi(g) > 0$. Let z be any arbitrary embedding of M into a Euclidean space, and let u be any arbitrary smooth mapping of M into a Euclidean space. Then, for any positive ε, $\varepsilon z \oplus u$ is an embedding. Since $\phi\ (f(\varepsilon z \oplus u)) \leqq 0$

for all $\varepsilon > 0$, it follows on letting $\varepsilon \to 0$ that $\phi(f(u)) \leqq 0$ for every smooth mapping of M into a Euclidean space (cf. formula (1) above). By formula (1) above, this is equivalent to the statement that $\phi(f(u)) \leqq 0$ for each smooth mapping of M into 1-dimensional Euclidean space, that is, that $\phi(f(u)) \leqq 0$ for each smooth real-valued function on M.

We now let $V \subseteq M$ be a coordinate patch, introduce coordinates $[x_1, x_2, \ldots, x_n] = [x_1, y] = x$ in V mapping V onto the unit sphere in Euclidean space and restrict the functional ϕ to the set S_V^∞ of tenors in S^∞ vanishing outside V. For $h \in S_V^\infty$, $\phi(h)$ may be written as $\phi(h) = \sum_{i,j=1}^{n} D^{ij}(h_{ij}(\cdot))$, where $D^{ij} = D^{ji}$ is a distribution defined in the unit sphere of n-dimensional Euclidean space, and where $h_{ij}(x)$ is the coordinate expression of the tensor $h \in S_V^\infty$. The above condition $\phi(f(n)) \leqq 0$ evidently implies that $\sum_{i,j=1}^{n} D^{ij}\left(\frac{\partial u}{\partial x_i}\frac{\partial u}{\partial x_i}\right) \leqq 0$ for all smooth functions u vanishing outside a subset of the unit sphere.

Let η be a smooth non-negative function in R^n, of total integral 1, vanishing outside the unit circle. We know from the general theory of distributions that the "convolutions" $D_\varepsilon^{ij}(y) = D^{ij}\left(\varepsilon^{-n}\eta\left(\frac{\cdot - y}{\varepsilon}\right)\right)$ are a family of C^∞ functions, defined in the sphere of radius $1 - \varepsilon$, and converging as $\varepsilon \to 0$ to D^{ij} in the sense of the theory of distributions. From the statement $\sum_{i,j=1}^{n} D^{ij}\left(\frac{\partial u}{\partial x_i}\frac{\partial u}{\partial x_j}\right) \leqq 0$, it is easily verified that

[*]
$$\sum_{i,j=1}^{n} \int D_\varepsilon^{ij}(x)\, \frac{\partial u(x)}{\partial x_i}\, \frac{\partial u(x)}{\partial x_j}\, dx \leqq 0$$

for each smooth function u vanishing outside the sphere of radius $1 - \varepsilon$. We shall show that [*] implies that

[**]
$$\sum_{i,j=1}^{n} D_\varepsilon^{ij}(x)\, \xi_i \xi_j \leqq 0 \quad \text{for each} \quad |x| < 1 - \varepsilon$$
$$\text{and each vector} \quad \xi \in R^n.$$

We proceed as follows. First note that, by the rotational symmetry and the homogeneity of the condition [*], it is sufficient to prove [**] in the special, notationally simpler case $\xi = [1, 0, \ldots, 0]$ and $x = [c, y]$, i.e., to prove that $D_\varepsilon^{11}(c, y) \leqq 0$ for each small c and $|y| < 1 - \varepsilon$. To establish this last inequality, let ψ be a smooth function of a real variable t, equal to a constant in a small neighborhood of $t = 0$ and vanishing for $|t| > \frac{1}{2}(1 - \varepsilon)$, and put $u_\delta(x) = \delta^{1/2}\psi\left(\frac{x_1}{\delta} - c\right)\psi(|y|)$. Then

$$\frac{\partial u_\delta}{\partial x_1}(x)\frac{\partial u_\delta}{\partial x_1}(x) = \delta^{-1}\left(\psi'\left(\frac{x_1}{\delta} - c\right)\right)^2 (\psi(|y|))^2,$$

so that

$$\int\left(\frac{\partial u_\delta}{\partial x_1}(x)\right)^2 dx = \iint (\psi'(x_1 - c))^2\,\psi(|y|)^2\,dx$$

is independent of δ, while all the other products of partial derivatives of $u_\delta(x)$ have integrals which go to zero as $\delta \to 0$. Thus, choosing ψ so that

$$\iint |\psi'(x_1 - c)|\,|\psi(|y|)|^2\,dx > 0,$$

putting u_δ for u in [*], and letting $\delta \to 0$, we find that

$$\int D_\varepsilon^{11}(c, y)\,(\psi(|y|))^2\,dy \leqq 0.$$

Therefore, letting ψ vary through a sequence of functions approaching a δ-function, we find that $D_\varepsilon^{11}(c, y) \leqq 0$ for $|y| < 1 - \varepsilon$. Therefore, as already observed, [**] follows.

Now note that if $-A$ is a positive symmetric matrix and B is a positive symmetric matrix, then $tr\,(AB) \leqq 0$. Indeed, we have

$$tr\,(AB) = tr\,(AB^{1/2}B^{1/2}) = -tr\,(B^{1/2}(-A)^{1/2}(-A)^{1/2}B^{1/2}) = -tr\,(CC^*),$$

where $C = B^{1/2}(-A)^{1/2}$. Since $CC^* \geqq 0$ we have $tr\,(CC^*) \geqq 0$ and our conclusion follows. Therefore it is a consequence of [**] that

[***] $$\sum_{i,j=1}^{n} D_\varepsilon^{ij}(x)\,h_{ij}(x) \leqq 0$$

for every smooth positive symmetric tensor $h_{ij}(x)$ vanishing outside $|x| < 1 - \varepsilon$. Integrating the inequality [***] and letting $\varepsilon \to 0$ we conclude that

$$\sum_{i,j=1}^{n} D^{ij}(h_{ij}) \leqq 0$$

for every positive symmetric tensor vanishing outside a compact subset of the unit sphere, i.e. that $\phi(h) \leqq 0$ for each positive $h \in S^\infty$ vanishing outside the coordinate neighborhood V. Since, by use of an appropriate partition of unity, any positive-definite $g \in S^\infty$ can be written as a sum of positive elements $h_i \in S^\infty$, each vanishing outside a certain coordinate patch of M, it follows that $\phi(g) \leqq 0$ for each $g \in K^\infty$. But this contradicts our original statement $\phi(g) > 0$, and thus completes the proof of assertion (ii).

Q.E.D.

2.4. Theorem: (Nash, [4]). For every compact Riemannian manifold M with a C^∞-metric, there exists a C^∞ isometric embedding of M in some Euclidean space.

Proof: If E^∞, K^∞ are the cones defined above, our theorem states simply that $E^\infty = K^\infty$. By the lemma above we know that E^∞ is dense in K^∞; and from (9) we also know that E^∞ has interior points.

Let $g \in K^\infty$, g_0 be an interior point of E^∞. As K^∞ is open, there exists an element $\bar{g} \neq g$ in K^∞ such that g is a convex combination of g_0 and g. But then $\bar{g}$ is also a cluster point of E^∞, because it belongs to K^∞. Moreover, g_0 is interior to E^∞ and E^∞ is convex. This implies that all points in the open segment joining g_0 and $\bar{g}$ (in particular g) are interior points of E^∞ (see [1], E.V.T., Chap. II, § 1, prop. 15), and we are done: $E^\infty = K^\infty$.

CHAPTER III

Degree Theory and Applications

A. A Form of Sard's Lemma

Our aim is to prove the following Theorem 3.1, which is related to Sard's lemma (cf. de Rham, "Variétés differentiables", Sec. 3, Th. 4).

3.1. Theorem: Let D be an open set in R^n, let f be a continuously differentiable mapping of D into R^n, and let $J(x)$ be the Jacobian determinant of f at x. Then for any measurable subset E of D the set $f(E)$ is measurable and

$$m\,(f(E)) \leqq \int_E |J(x)|\,dx.$$

We begin by recalling a few definitions. For any point v_0 of R^n and any set of n linearly independent vectors $a_1, \ldots, a_n$ in R^n, the *parallelotope P with initial vertex v_0 and edgevectors $a_1, \ldots, a_n$* is the set of all points of R^n of form $x = v_0 + \sum_{i=1}^{n} \lambda_i a_i$, where $\lambda_1, \ldots, \lambda_n$ are real numbers such that the $0 \leqq \lambda_i \leqq 1$, $i = 1, \ldots, n$. The point $v_0 + \frac{1}{2}\sum_{i=1}^{n} a_i$ is called the *center* of the parallelotope.

For fixed k the set of those points of P for which λ_k has a fixed value equal to either 0 or 1 is called an $(n - 1)$-*dimensional face* (or, briefly, *face*) of P, so that the number of faces of P is $2n$. The point $v_0 + \lambda_k a_k + \frac{1}{2}\sum_{\substack{i=1 \\ i \neq k}}^{n} a_i$ is called the center of the face.

It is immediate that the parallelotope P with initial vertex at the origin and edge-vectors $a_1, \ldots, a_n$ is the image of the unit cube

$$C = \{x = (x^1, \ldots, x^n) : 0 \leqq x^i \leqq 1, \quad i = 1, \ldots, n\}$$

under the non-singular linear transformation $h : R^n \to R^n$ given by

$$h(x) = h(x^1, \ldots, x^n) = \sum_{i=1}^{n} x^i a_i;$$

moreover the image of the unit cube by any non-singular linear transformation R^n onto itself is a parallelotope of this form. It follows that P is compact, and that the frontier of P is the union of the $2n$ faces of P. Moreover, (see, for example, Zaanen [4], p. 160) the n-dimensional measure of P is equal to $|\det(h)| = \det(a_i^j)|$, where a_i is the j-th coordinate of a_i; and obviously these last results extend to a parallelotope with any initial vertex.

Throughout the following discussion we use the ordinary Euclidean norm for points of R^n and the corresponding norm for linear transformations of R^n into itself, and we denote the inner product of x and y by $x \cdot y$. We use $A(n)$ to denote a positive constant depending only on n, not necessarily the same on any two occurrences.

For our proof of Theorem 3.1 we require two simple geometrical inequalities, which we state below.

1. *Let F be a set in R^n contained in a hyperplane H, let x_0 be a fixed point of F, and let $\|x - x_0\| \leqq d$ whenever $x \in F$. Let also G be the set of points of R^n whose distance from F is less than δ. Then G is measurable* (*since it is open*) *and*

$$m(G) \leqq 2^n (d + \delta)^{n-1} \delta . \tag{α.1}$$

It is evident that G lies between the two hyperplanes parallel to H and distance δ from it, and to prove (α.1) we construct a parallelotope containing G with two of its faces in these hyperplanes.

By a suitable translation we may suppose that H contains the origin, so that H is an $(n-1)$-dimensional vector subspace of R^n. We can therefore find a unit vector a_1 such that $x \cdot a_1 = 0$ for all $x \in H$ (i.e. such that a_1 is orthogonal to H), and then we can find vectors $a_2, \dots, a_n$ such that $\{a_1, a_2, \dots, a_n\}$ is a complete orthogonal set in R^n. Let now $y \in G$. Since every vector in R^n can be expressed as a linear combination of the a_i, there exist real numbers $\lambda_1, \dots, \lambda_n$ such that

$$y - x_0 = \sum_{i=1}^{n} \lambda_i a_i .$$

Further, since the distance of y from F is less than δ, there exists $x \in F$ (possibly identical with y) such that $\|y - x\| < \delta$, and then writing

$$y - x = (y - x_0) - (x - x_0),$$

we obtain

$$\lambda_1 = (y - x_0) \cdot a_1 = (y - x) \cdot a_1 - (x - x_0) \cdot a_1 = (y - x) \cdot a_1,$$

whence

$$|\lambda_1| \leqq \|y - x\| \, \|a_1\| = \|y - x\| < \delta .$$

Also

$$\|y - x_0\| \leqq \|y - x\| + \|x - x_0\| < \delta + d,$$

so that for $i = 2, \dots, n$,

$$|\lambda_i| = |(y - x_0) \cdot a_i| \leqq \|y - x_0\| \, \|a_i\| = \|y - x_0\| \leqq d + \delta .$$

It follows that G is contained in the (fixed) parallelotope with center x_0 and edge-vectors $2\delta a_1$, $2(d + \delta)\, a_i$, $i = 2, \dots, n$, and since the measure of this parallelotope is $2^n (d + \delta)^{n-1} \delta \, |\det (a_i^j)| = 2^n (d + \delta)^{n-1} \delta$, the result follows.

2. *Let h be a linear transformation of R^n into itself, let P be the image by h of the unit cube $C = \{x = (x^1, \dots, x^n)\colon 0 \leqq x^i \leqq 1,\ i = 1, \dots, n\}$, and let Q be the set of points of R^n whose distance from P is less than δ. Then Q is measurable (since it is open) and*

$$m(Q) \leqq |\det (h)| + A(n) \, (\|h\| + \delta)^{n-1} \, \delta .$$

Suppose first that $\det (h) = 0$, so that h is singular. In this case P is contained in a hyperplane, and we apply Lemma 1 to $F = P$, taking x_0 to be the image of the center w_0 of C. Since

$$\|h(w) - h(w_0)\| = \|h(w - w_0)\| \leqq \|h\| \, \|w - w_0\| \leqq \tfrac{1}{2}\sqrt{n}\, \|h\|$$

whenever $w \in C$, we have $\|x - x_0\| \leqq \tfrac{1}{2}\sqrt{n}\, \|h\|$ whenever $x \in P$, whence statement 1 above gives

$$m(Q) \leqq 2^n(\tfrac{1}{2}\sqrt{n}\, \|h\| + \delta)^{n-1}\, \delta \leqq A(n)\, (\|h\| + \delta)^{n-1}\, \delta,$$

as required.

Suppose next that $\det(h) \neq 0$. In this case P is a parallelotope with measure $m(P) = |\det(h)|$, and it is therefore enough to prove that the open set $Q - P$ has measure not exceeding $A(n)\,(\|h\| + \delta)^{n-1}\delta$. Since P is compact, for each $y \in Q - P$ there exists $x \in P$ such that $\|y - x\|$ is equal to the distance of y from P, and evidently x is a frontier point of P, so that x lies on one or more $(n - 1)$-dimensional faces of P. Since P and $2n$ such faces and each face is the image by h of a face of C, it is now enough to prove that if B is a face of C and E is the set of points of R^n whose distance from $h(B)$ is less than δ, then

$$m(E) \leqq A(n)\, (\|h\| + \delta)^{n-1}\, \delta. \tag{α.2}$$

To prove this last result we observe that $h(B)$ is contained in a hyperplane, so that we can apply 1 to $F = h(B)$. We choose x_0 to be the center of the face $h(B)$ of P, so that $\|x - x_0\| \leqq \tfrac{1}{2}\sqrt{(n-1)}\, \|h\|$ whenever $x \in h(B)$, and then 1 gives

$$m(E) \leqq 2^n(\tfrac{1}{2}\sqrt{(n-1)}\, \|h\| + \delta)^{n-1}\, \delta \leqq A(n)\, (\|h\| + \delta)^{n-1}\, \delta.$$

This proved (α.2), and completes the proof of 2.*

From 2 we deduce immediately:

3. *Let C be a closed cube in R^n with sides parallel to the axes and of length α, let h be a linear transformation of R^n into itself, and let Q be the set of points of R^n whose distance from the set $h(C)$ is less than $\alpha\delta$. Then Q is measurable (since it is open) and*

$$m(Q) \leqq m(C)\, \{|\det(h)| + A(n)\, (\|h\| + \delta)^{n-1}\, \delta\}. \tag{α.3}$$

---

* In the case in which $\det(h) \neq 0$ it is tempting to estimate $m(Q)$ by using the inequality $m(Q) \leq m(P')$, where P' is the smallest parallelotope containing Q with sides parallel to those of P, but unfortunately the measure $m(P')$ tends to infinity as we approach the singular case, i.e. as $\det(h)$ tends to 0 (this is easily seen from a diagram illustrating the plane case). Most proofs of the change of variable formula in which the estimate of the measure of a parallelotope appears do in fact use an estimate of the form $m(Q) \leq m(P')$, and it is for this reason that the hypothesis $\inf |J(x)| > 0$ is essential to such proofs.

By applying 3 to the derivative of a differentiable mapping, we obtain the following result; in this we use the definition of derivative given in 1.9.

4. *Let C be a closed cube in R^n with center x_0 and with sides parallel to the axes, let f be a differentiable mapping of C into R^n, and let $J(x)$ be the Jacobian determinant of f at x. Then*

$$(\alpha.4) \qquad m^*(f(C)) \leqq m(C)\{|J(x_0)| + A(n)(\|f'(x_0)\| + \eta)^{n-1}\eta\},$$

where $\eta = \sup_{x \in C} \|f'(x) - f'(x_0)\|$ *and m^* denotes outer Lebesgue measure.*

To prove (α.4) let α be the length of the sides of C, and let P be the image of C by the linear transformation $f'(x_0): R^n \to R^n$. By the mean value theorem applied to the function $f - f'(x_0)$ (cf. the proof of Corollary 1.45), we have for each x of C

$$\|f(x) - f(x_0) - f'(x_0)(x - x_0)\| \leqq \eta\,\|x - x_0\| < \eta\alpha\sqrt{n},$$

and this inequality expresses the fact that the point $f(x) - f(x_0) + f'(x_0)(x_0)$ of the translate $f(C) - f(x_0) + f'(x_0)(x_0)$ of $f(C)$ is at a distance less than $\eta\alpha\sqrt{n}$ from the point $f'(x_0)(x)$ of P. It follows that this translate of $f(C)$ is contained in the set of points of R^n whose distance from P is less than $\eta\alpha\sqrt{n}$, and applying 3 (and noting that $\det(f'(x_0)) = J(x_0)$) we immediately obtain the inequality (α.3).

5. *Let D be an open set in R^n, let f be a continuously differentiable mapping of D into R^n, and let $J(x)$ be the Jacobian determinant of f at x. Then for any measurable subset E of D*

$$(\beta.1) \qquad m^*(f(E)) \leqq \int_E |J(x)|\,dx,$$

where m^* denotes outer Lebesgue measure.

Suppose first that E is a closed cube C with sides parallel to the axes. Since f' is continuous on C, we can divide C into a finite number of non-overlapping closed cubes $C_1, \ldots, C_N$ with centers $x_1, \ldots, x_N$ and with sides parallel to the axes such that $\|f'(x) - f'(x_k)\| \leqq \varepsilon$ whenever $x \in C_k$ ($k = 1, \ldots, N$). By 4, for each cube C_k we have

$$m^*(f(C_k)) < m(C_k)\{|J(x_k)| + A\varepsilon\},$$

where A is independent of k, so that also

$$m^*(f(C)) \leqq \Sigma\, m^*(f(C_k)) < \Sigma\, |J(x_k)|\, m(C_k) + A\varepsilon m(C),$$

the summations being extended over all cubes C_k. When the maximum diameter of the cubes C_k tends to 0 the sum $\Sigma\ |J(x_k)|\ m(C_k)$ tends to the Riemann integral of $|J(x)|$ over C, and since ε is arbitrary we therefore obtain

$$m^*(f(C)) \leqq \int_C |J(x)|\ dx, \tag{β.2}$$

which is (β.1) for $E = C$.

Suppose next that E is a measurable subset of D. Then we can find a set E_1 containing E and with measure equal to that of E such that E_1 is the intersection of a contracting sequence of open sets $O_n \subset D$. If now C is a closed cube contained in D with sides parallel to the axes, then for each fixed n the set $C \cap O_n$ is a countable union of non-overlapping closed cubes with sides parallel to the axes, and applying (β.2) to each such cube and summing we obtain

$$m^*(f(C \cap O_n)) \leqq \int_{C \cap O_n} |J(x)|\ dx,$$

whence also

$$m^*(f(C \cap E)) \leqq \int_{C \cap O_n} |J(x)|\ dx \tag{β.3}$$

(since $E \subset O_n$). Since J is bounded above on C, the integral on the right of (β.3) is finite, and so tends to $\int_{C \cap E_1} |J(x)|\ dx$ as n tends to $+\infty$, whence

$$m^*\,(f(C \cap E)) \leqq \int_{C \cap E_1} |J(x)|\ dx = \int_{C \cap E} |J(x)|\ dx.$$

Since D is a countable union of non-overlapping cubes such as C, the general result (β.1) follows.

6. *Let D be an open set contained in R^n, and let f be a continuously differentiable mapping of D into R^n. Then $f(E)$ is measurable for every measurable set $E \subset D$.*

(For a proof of this under more general hypotheses see Rado and Reichelderfer [1], pp. 337, 214).

Let $J(x)$ be the Jacobian determinant of f at x. It follows immediately from 5 that if E_0 is the subset of D where $J(x) = 0$, then $f(E_0)$ has measure 0, so that $m\,(f(E \cap E_0)) = 0$ for every measurable $E \subset D$. Since $D - E_0$ is open, it is therefore enough to prove the above result when $J(x) \neq 0$ on D.

Suppose then that $J(x) \neq 0$ on D, so that f is locally a homeomorphism. The open set D is a countable union of closed cubes, and, by the Heine-Borel theorem, we can cover each of these cubes with a finite number of closed cubes on each of which f is a homeomorphism. Hence D is a countable union of closed cubes C_i on each of which f is a homeomorphism, and since $f(E) = \bigcup f(E \cap C_i)$, it is enough to prove that $f(E \cap C)$ is measurable whenever E is measurable and C is a closed cube in D on which f is a homeomorphism.

If E is closed, so are $E \cap C$ and $f(E \cap C)$, and hence if E is a countable union of closed sets, then $f(E \cap C)$ is measurable. Since any measurable set is the union of a set of measure zero and a set which is a countable union of closed sets, it is now enough to prove that $f(E \cap C)$ is measurable when E is of measure zero, and this follows immediately from 5. This completes the proof of 6, and hence also of Theorem 3.1.

References

1. T. Rado and P. V. Reichelderfer, *Continuous transformation in analysis* (Berlin, 1955).
2. G. de Rham, *Variétés différentiables* (Paris, 1955).
3. J. Schwartz, "The formula for change in variables in a multiple integral", Amer. Math. Monthly **61** (1954), 81–5.
4. A. C. Zaanen, *An introduction to the theory of integration* (Amsterdam, 1958).

B. Definition of the Degree of a C^1 Mapping in R^n

Notation: Until further comment, D will denote an open bounded set of R^n, the Euclidean space whose coordinates are $x = (x_1, \ldots, x_n)$. Let ∂D denote the boundary of D.

Most of the mappings appearing below are continuous on $\bar{D}$. We shall write C for the space $C(\bar{D}, R^n)$ of continuous mappings defined on $\bar{D}$ and having values in R^n. By a C^r function on $\bar{D}$ we mean a function having derivatives on a neighborhood of $\bar{D}$ up to order r which coincide with restrictions of continuous functions on $\bar{D}$. Cf. Chapter I for the topology of C^r.

Suppose that $\phi \in C$ is C^1 on D and that $p \in R^n$ is a point not belonging to $\phi(\partial D)$. We shall define the degree of ϕ with respect to p and D; it will be an integer denoted by $\deg(p, \phi, D)$.

If $Z \subset D$ is the set of critical points of ϕ, i.e. points at which J_ϕ, the Jacobian of ϕ vanishes, and $\phi^{-1}(p) \cap Z = \varnothing$, then the set $\phi^{-1}(p)$ is discrete, by the implicit function theorem; since $\bar{D}$ is compact, this set is *finite*.

At each $x \in \phi^{-1}(p)$, J_ϕ does not vanish. Then its sign is unambiguously defined and we define

$$\text{(II.1)} \qquad \deg(p, \phi, D) = \sum_{\phi(x)=p} \operatorname{sign} J_\phi(x).$$

Suppose now that $\phi^{-1}(p) \cap Z \neq \varnothing$.

By Sard's Lemma 3.1, $\phi(Z)$ has measure zero in R^n, and in particular, has empty interior. This implies that the point p may be approximated as closely as desired by points q for which $\phi^{-1}(q) \cap Z = \varnothing$. For each q, the degree is defined as above. Then, by definition, the degree of p is

$$\text{(II.2)} \qquad \deg(p, \phi, D) = \lim_{\substack{q \to p \\ \phi^{-1}(q) \cap Z = \varnothing}} \deg(q, \phi, D).$$

In order to justify this definition we must prove that the limit exists (and that it is independent of the choice of the q's). This will follow from more general results to be obtained later (see Corollary 3.10). As a first remark we note that if $p \notin \phi(\bar{D})$, then $\deg(p, \phi, D) = 0$.

Let us consider the special case when $\phi^{-1}(p) \cap Z = \varnothing$. Suppose that $\{f_\varepsilon\}$ is a family of continuous functions $f_\varepsilon : R^n \to R$ with the properties

(II.3)

(i) $$\int_{R^n} f_\varepsilon(x)\, dx = 1$$

(ii) K_ε = support f_ε = sphere of radius ε and center at p.

Let us consider the integrals

$$I_\varepsilon = \int_{\bar{D}} f_\varepsilon(\phi(x))\, J_\phi(x)\, dx.$$

We shall prove that $I_\varepsilon = \deg(p, \phi, D)$ for every ε small enough.

Let $p^1, \ldots, p^s$ be the elements of $\phi^{-1}(p)$ (as was observed, this set is finite).

When ε is small enough, there exist neighborhoods $A_\varepsilon^1, \ldots, A_\varepsilon^s$ of $p^1, \ldots, p^s$ such that ϕ naps each A^i homeomorphically onto K_ε, the sphere of radius ε around p. Observe that $f_\varepsilon(\phi(x))$ vanishes outside $\bigcup_{i=1}^{s} A_\varepsilon^i$. It follows that

$$\text{(1)} \qquad \int_D f_\varepsilon(\phi(x))\, J_\phi(x)\, dx = \sum_{i=1}^{s} \int_{A_\varepsilon^i} f_\varepsilon(\phi(x))\, J_\phi(x)\, dx.$$

Since $J_\phi(p^i) \neq 0$, for $i = 1, \ldots, s$ (recall that $\phi^{-1}(p) \cap Z = \varnothing$), by choosing ε small enough, we may assume that $J_\phi \neq 0$ at every point in every A_ε^i. In that case sign J_ϕ is (defined and) constant on every A^i_ε. Hence we may consider $\phi : A^i_\varepsilon \to K_\varepsilon$ as a change of variables and apply the classical theorems on change of variables in an integral. This leads to the equality:

$$(2) \qquad \int_{A^i_\varepsilon} f_\varepsilon(\phi(x))\, J_\phi(x)\, dx = \operatorname{sign} J_\phi(p^i) \int_{K_\varepsilon} f_\varepsilon(x)\, dx = \operatorname{sign} J_\phi(p^i).$$

Combining (1) and (2) yields

$$\int_D f_\varepsilon(\phi(x))\, J_\phi(x)\, dx = \Sigma \operatorname{sign} J_\phi(p^i), \quad \{p^i\} = \phi^{-1}(p),$$

i.e.,

$$(\text{II}.4) \qquad \deg(p, \phi, D) = \int_D f_\varepsilon(\phi(x))\, J_\phi(x)\, dx$$

for every family of functions with the properties (1.3) and provided ε is small enough.

This expression may be adopted as an alternate definition of deg (p, ϕ, D) whenever $\phi^{-1}(p) \cap Z = \varnothing$ (see E. Heinz, [1]). We shall prove some properties of the degree as so expressed. First we need some lemmas.

C. Some Functions are Divergences

3.2. Lemma: Suppose v is a C^1 vector function: $(v^1, \ldots, v^n) = v : E^n \to E^n$, and that $f = \operatorname{div} v \left(= \sum_i \dfrac{\partial v^i}{\partial x_i} \right)$. If v vanishes outside a bounded set K, then

$$\int_{R^n} f(x)\, dx = 0.$$

Proof: Compute.

Suppose now that $\phi \in C$ and ϕ in C^2 on D, and that v is as in Lemma 3.2. K is the (compact) *support* of f.

3.3. Lemma: If $K \cap \phi(\partial D) = \varnothing$, the function

$$g(x) = f(\phi(x))\, J_\phi(x)$$

is the divergence of some vector valued C^1 function u with support included in D.

Proof: Let $a^{k,l}$ be the minor determinant corresponding to the (k, l)-th entry in the Jacobian matrix $\dfrac{\partial \phi^i}{\partial x_j}$ and define u by

$$u^i(x) = \sum_j v^j(\phi(x))\, a^{i,j}(x), \qquad (i = 1, 2, \ldots, n).$$

Then u is C^1 since ϕ is C^2 and v is C^1. From the hypothesis $\phi(\partial D) \cap K = \varnothing$ it also follows that u may be considered as defined on all of R^n (with the value zero outside D) and that its support is then included in D, as required. It only remains to prove that $\operatorname{div} u = g$. We simply compute as follows:

$$u_i^i(x) = \sum_{k,j} [v_k^j(\phi(x))\phi_i^k(x)\, a^{i,j}(x) + v^j(\phi(x))\, a_i^{i,j}(x)],$$

$$\begin{aligned}\operatorname{div} u &= \sum_i u_i^i = \operatorname{div} v\, J_\phi(x) + \sum_{i,j} v^j(\phi(x))\, a_i^{i,j}(x)\\ &= g(x) + \sum_j v^j(\phi(x)) \Big(\sum_i a_i^{i,j}(x)\Big).\end{aligned}$$

By definition we have

$$(n-1)!\, a^{i,j} = \sum \varepsilon_{j,j_1,\ldots,j_{n-1}}^{i,m_1,\ldots,m_{n-1}} \phi_{m_1}^{j_1} \cdots \phi_{m_{n-1}}^{j_{n-1}}$$

(where ε means sign of the permutation $m_h \to j_h$), and then, differentiating, the antisymmetry leads to $\sum_i a_i^{i,j} = 0$, which implies the desired result.

3.4. Lemma: Let f be a continuous function defined on R^n having support K contained in D. Let $x^0 \in R^n$ and suppose that the convex hull of $K \cup (K - x^0)$ (where $K - x^0$ denotes the set obtained from K by the translation induced by $-x^0$) is contained in D. Then the function

$$f(x) - f(x + x^0)$$

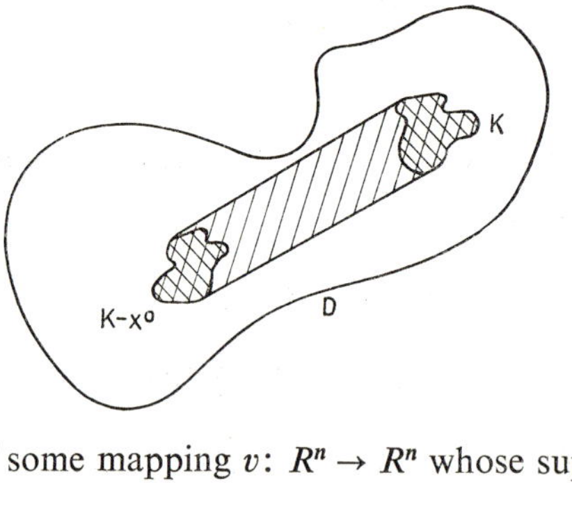

is the divergence of some mapping v: $R^n \to R^n$ whose support is contained in D.

Proof: Let $\phi(x) = f(x) - f(x + x^0)$. Clearly ϕ has support equal to $K \cup (K - x^0)$.

Now define

$$\Phi(x) = \int_{-\infty}^{0} \phi(x + tx^0)\, dt,$$

$$v^i(x) = x_i^0 \Phi(x).$$

It is easily verified that v^i has support contained in the convex hull of $K \cup (K - x^0)$. Moreover, if $v = (v^1, \ldots, v^n)$, we claim that div $v = \phi$. In fact, we have

$$\text{div } v = \sum_i v_i^i = \sum_i x_i^0 \frac{\partial \Phi}{\partial x_i}.$$

But the last term is the directional derivative of Φ in the direction x^0, and consequently equals

$$\left. \frac{d}{dt} \Phi\,(x + tx^0) \right|_{t=0}.$$

By definition of Φ, it follows that

$$\left. \frac{d}{dt} \Phi\,(x + tx^0) \right|_{t=0} = \left. \frac{d}{dt} \left(\int_{-\infty}^{0} \phi\,(x + (t + u)\, x^0)\, du \right) \right|_{t=0}$$

$$= \left. \int_{-\infty}^{0} \frac{d}{dt} \phi\,(x + (t + u)\, x^0)\, du \right|_{t=0}$$

$$= \int_{-\infty}^{0} \left(\frac{d}{du} \phi\,(x + ux^0) \right) du = \phi(x),$$

and then div $v(x) = \phi(x) = f(x) - f(x + x^0)$ as desired.

3.5. Corollary: Let $x(s)$ be a continuous curve in R^n, $0 \leqq s \leqq 1$, and let f be continuous $f\colon R^n \to R$ with support K contained in D. Suppose

(i) K is contained in a convex compact set M contained in D.
(ii) $M - x(s)$ never touches the boundary of D.

Then $f(x + x(0)) - f(x + x(1))$ is the divergence of some C^1 mapping v: $R^n \to R^n$ whose support in contained in D.

Proof: Let us define an equivalence relation on the set of values of s by $s_1 \sim s_2$ iff $f(x + x(s_1)) - f(x + x(s_2))$ is the divergence of a mapping with support in D. Lemma 3.4 allows us to conclude that every class modulo $\sim$ is open. In fact, for every s, $M + x(s)$ being a compact *convex* subset of D there exists an open convex neighborhood of $M + x(s)$ also contained in D. By the continuity of $x(s)$, the convex hull of $(M + x(s)) \cup (M + x(s'))$ is contained in such a neighborhood, when s' is close to s. Then Lemma 3.4 applies and $s \sim s'$. Thus we conclude that every class is open. But the connectedness of $[0, 1]$ therefore implies that there is only one class. This means that $0 \sim 1$ or that $f(x + x(0)) - f(x + x(1))$ is a divergence, and we are done.

D. Back to the Definition

We begin by proving a lemma.

3.6. Lemma: Let $\phi \in C$ and be C^2 on D. Consider two points $p_1, p_2 \in R^n - \phi(\partial D)$ and such that $\phi^{-1}(p_1) \cap Z = \phi^{-1}(p_2) \cap Z = \varnothing$. Then if p_1 and p_2 belong to the same component of the open set R^n, we have

$$\deg(p_1, \phi, D) = \deg(p_2, \phi, D).$$

Proof: Since $\phi^{-1}(p_1) \cap Z = \varnothing$, we know that $\deg(p, \phi, D)$ may be computed by means of a family of functions with the simple properties described in II.3 as follows:

$$\deg(p_1, \phi, D) = \int_D f_\varepsilon(\phi(x))\, J_\phi(x)\, dx$$

for ε small.

If we suppose that p_2 lies in the same component of $R^n - \phi(\partial D)$ as does p_1, then there exists a continuous curve $x(s)$, $0 \leqq s \leqq 1$ such that $x(0) = 0$, $x(1) = p_2 - p_1$ and $x(s) + p_1$ lies in that component. Since $x(s)$ is compact there exists $\varepsilon > 0$ such that if K_ε is the ε-sphere around p_1, $K_\varepsilon + x(s)$ never touches the boundary of $R^n - \phi(\partial D)$. Then Lemma 3.4 may be applied and yields the conclusion that

$$f_\varepsilon(x) - f_\varepsilon(x + p_2 - p_1)$$

is a divergence. Therefore by Lemma 3.3,

$$f_\varepsilon(\phi(x))\, J_\phi(x) - f_\varepsilon(\phi(x) + p_2 - p_1)\, J_\phi(x)$$

is also the divergence of a mapping with support in D. In such a case, Lemma 3.2 implies that

$$\text{(a)} \qquad \int_D f_\varepsilon(\phi(x))\, J_\phi(x)\, dx = \int_D f_\varepsilon(\phi(x) + p_2 - p_1)\, J_\phi(x)\, dx$$

or, according to formula (II.4):

(b) $$\deg(p_1, \phi, D) = \int_D f_\varepsilon(\phi(x) + p_2 - p_1)\, J_\phi(x)\, dx.$$

But now the functions $g_\varepsilon(x) = f_\varepsilon(x + p_2 - p_1)$, $\varepsilon > 0$ have the properties (II.3) around p_2, and consequently

$$\deg(p_2, \phi, D) = \int_D g_\varepsilon(\phi(x))\, J_\phi(x)\, dx$$

$$= \int_D f_\varepsilon(\phi(x) + p_2 - p_1)\, J_\phi(x)\, dx.$$

Formulas (a), (b) and the last one together imply

$$\deg(p_1, \phi, D) = \deg(p_2, \phi, D)$$

and we are done.

Consequences: Suppose that $p \in R^n$ but $\phi^{-1}(p) \cap Z = \varnothing$. In that case, for every q sufficiently close to p, q belongs to some well-defined component of $R^n - \phi(\partial D)$, namely that containing p. But then by Lemma 3.6, $\deg(q, \phi, D)$ is constant when q is near p and $\phi^{-1}(q) \cap Z = \varnothing$. This justifies the definition (II.2) when ϕ is C^2. Taken together Lemma 3.6 and the definition (II.2) imply:

3.7. Corollary: Let $\phi \in C$ and be C^2 on D. Then

$$\deg(p, \phi, D)$$

is constant on every component of $R^n - \phi(\partial D)$.

This corollary will also be true for continuous mappings after we define the degree for such mappings. The unnatural hypothesis that ϕ is C^2 needs to be eliminated first, we will do so in the first corollary of the next lemma.

3.8. Lemma: Let $\phi \in C$ and be C^1 on D. Then for each $p \notin \phi(\partial D) \cup \phi(Z)$, there exists a C^1 neighborhood U of ϕ such that for every $\psi \in U$ we have $p \notin \psi(\partial D)$ and

$$\deg(p, \phi, D) = \deg(p, \psi, D).$$

Proof: Let y_j, $j = 1, \ldots, k$ be the elements of the finite set $\phi^{-1}(p)$. Let B_j be an open ball around y_j, whose radius r_j will be determined below.

First of all, if each r_j is small enough, the family $\{B_j\}$ is disjoint. Our aim now is to prove that there exists a C^1 neighborhood U such that for every ψ in U, the equation $\psi(x) = p$ has one and only one solution in each B_j and no others.

The hypothesis $p \notin \phi(Z)$ implies that the derivative ϕ' of ϕ (the Jacobian matrix) has an inverse at each y_j. By decreasing the radii r_j it is possible to guarantee that ϕ' has an inverse at every point of each B_j and moreover that:

$$|(\phi'(y_j))^{-1}(\phi'(y) - \phi'(y_j))| < \tfrac{1}{2}, \qquad y \in B_j \tag{1}$$

where the norm $|\ |$ stands for the norm for operators on R^n into itself.

Let us suppose that this is the case for radii $r_1, \ldots, r_k$ and let r be the smallest of these. If we let F be the set $D - \bigcup_j B_j$, then F is compact and $p \notin \phi(F)$. Finally, let a be a positive number such that $|(\phi'(y_j))^{-1}| > a, j = 1, \ldots, k$.

Now we are able to define the C^1 neighborhood of ϕ that will give a desired solution.

Choose U to be a ball (in the C^1 sense) such that for every $\psi \in U$ the following holds:

(a) $p \notin \psi(F)$.
(b) $\psi'(y)$ is invertible when $u \in \bigcup B_j$, and $|(\psi'(y_k))|^{-1} > a$.
(c) $|(\psi'(y_j))^{-1}(\psi'(y) - \psi'(y_j))| < \frac{3}{4}$, when $y \in B_j$ and for $j = 1, \ldots, k$.
(d) $|\phi(x) - \psi(x)| < \dfrac{ar}{4}$ for all $x \in \bar{D}$.

We remark that each one of the properties (a), (b), (c), (d) *defines* an open set in the C^1 sense (call these sets A, B, C, D), that the sets A, B, D obviously contain ϕ and that $\phi \in C$ by (1).

Therefore U may be chosen as any C^1 ball around ϕ contained in $A \cap B \cap C \cap D$.

Observe that since ϕ' is invertible over each B_j, the sign of the Jacobian determinant is constant. U being a ball, it is connected. Then the same result follows for every ϕ in U and clearly

$$\begin{aligned} &\operatorname{sign} J_\psi(y) = \operatorname{sign} J_\phi(y_j) \\ &\text{when} \quad j \in B_j \quad \text{and} \quad \psi \in U. \end{aligned} \tag{2}$$

We now show that properties (a), (b), (c) and (d) imply that the function ψ has one and only one root of the equation $\psi(x) = p$ in each B_j. Of course property (a) tells us that there are no roots outside $\bigcup B_j$ (remember that $F = \bar{D} - \bigcup B_j$). The only problem is to see what happens in one B_j, say, in B_1.

Define

$$\eta(y) = (\psi'(y_1))^{-1} (\psi(y)).$$

From (c) it follows that

$$|\eta'(y) - I| < \tfrac{3}{4}, \qquad y \in B_1.$$

Then by Corollary 1.19 it follows that η is one-to-one on B_j. This implies that ψ is one-to-one on B_1 and therefore it has *at most* one root there.

But from the same corollary we also conclude that $\eta(B_1)$ covers a ball B of radius $(1 - \frac{3}{4}) r_1 = \frac{1}{4}r_1 > \frac{1}{4}r$ around $\eta(y_1)$.

As $|(\psi'(y_1))^{-1}| > a$, it follows that $\psi'(y_1)(B)$ covers a ball V of radius $a \cdot \frac{1}{4}r$ around $\psi'(y_1)(\eta(y_1))$. But $\psi(y) = (\psi'(y_1))^{-1} (\eta(Y))$ and then $\psi(B_1)$ covers V. This means that every point $x \in R^n$ for which $|x - \psi(y_1)| < \frac{1}{4}ra$ is of the form $x = \psi(b)$, $b \in B_1$. But by (d), $|p - \psi(y_1)| = |\phi(y_1) - \psi(y_1)| < \dfrac{ar}{4}$, and then the equation $\psi(x) = p$ has *at least* one solution in B_1. The same holds for every B_j and we have $\psi^{-1}(p) \cap B_j = \{\bar{y}_j\}, j = 1, \ldots, k$. But then, by (2)

$$\deg(p, \psi, D) = \sum_j \text{Sign } J_\psi(\bar{y}_j) = \sum \text{Sign } J_\phi(y_j) = \deg(p, \phi, D).$$

Q.E.D.

3.9. Corollary: Let $\phi \in C(\bar{D})$ and be C^1 on D. If p, q do not belong to $\phi(\partial D) \cup \phi(Z)$ and belong to the same component of $R^n - \phi(\partial D)$, then

$$\deg(p, \phi, D) = \deg(q, \phi, D).$$

Proof: Choose a continuous curve $x(s)$ joining $p = x(0)$ and $q = x(1)$. Since $x(s)$ is compact and disjoint from $\phi(\partial D)$, when ψ is C^0 near ϕ, $x(s)$ is also disjoint from $\psi(\partial D)$. This means that p and q belong to the same component of $R^n - \psi(\partial D)$. Therefore if we choose ψ to be C^2,

$$\deg(p, \psi, D) = \deg(q, \psi, D).$$

This holds for every ψ of class C^2 and C^0 near ϕ. But since $\deg(p, \psi, D) \to \deg(p, \phi, D)$ and $\deg(q, \psi, D) \to \deg(q, \phi, D)$ as $\psi \to \phi$ in the C^1 sense, we conclude that

$$\deg(p, \phi, S) = \deg(q, \phi, D).$$

Q.E.D.

3.10. Corollary: Let p not belong to $\phi\,(\partial D)$. The expression

$$\deg\,(q, \phi, D)$$

has a limit when $q_j \to p$ and the q_j's belong to $R^n - (\phi\,(\partial D) \cup \phi(Z))$.

Proof: Obvious by the previous lemma.

The conclusion is that the degree of ϕ has a meaning for every point $p \notin \phi\,(\partial D)$ in the sense of the definition (II.1, II.2) which may therefore be adopted for every C^1 mapping.

The next step is to remove the condition $\phi \in C^2$ in Corollary 3.7.

3.11. Proposition: Let $\phi \in C$ and be C^1 on D. Then the degree

$$\deg\,(p, \phi, D)$$

as defined in (II.1) (II.2) is constant on every component of $R^n - \phi\,(\partial D)$.

Proof: This follows from 3.9 and the fact that $\phi(Z)$ has empty interior.

Next we remove the condition $p \notin \phi(Z)$ in 3.8.

3.12. Proposition: Let $\phi \in C$ and be C^1, and p be a point not in $\phi\,(\partial D)$. In that case there exists a C^1 neighborhood U of ϕ such that for every ψ in U, $p \notin \psi\,(\partial D)$ and:

$$\deg\,(p, \psi, D) = \deg\,(p, \phi, D).$$

Proof: This follows from 3.9 and the fact that $\phi(Z)$ has empty interior.

3.13. Corollary: Let $\{\phi_t\}$ be a family of C^1 mappings in C depending continuously in the C^1 sense on the real parameter t, $0 \leqq t \leqq 1$. If $p \notin \phi_t\,(\partial D)$ for every t, then

$$\deg\,(p, \phi_0, D) = \deg\,(p, \phi_1, D).$$

Proof: Essentially the same argument as in the proof of Corollary 3.5: the equality $\deg\,(p, \phi_t, D) = \deg\,(p, \phi_u, D)$ defines an equivalence relation $t \sim u$. Proposition 3.12 implies that each class is open and the connectedness of $[0, 1]$ allows us to conclude that $0 \sim 1$. This is the claim.

E. The Continuous Case

We are approaching the most important point in this chapter: the definition of the degree for *every continuous mapping*. This will demonstrate the topological character of this concept of degree. The definition is as follows:

3.14. Definition: Let $\phi \in C$ and let $\{\phi_n\}$ be a sequence of functions of class C^1 converging uniformly to ϕ. Then for every $p \notin \phi\ (\partial D)$ a sequence $\deg(p, \phi_n, D)$, $n \geqq N$ is defined, the limit

$$\lim_{n\to\infty} \deg(p, \phi_n, D)$$

exists and does not depend on $\{\phi_n\}$, and we then define

$$\deg(p, \phi, D) = \lim_{n\to\infty} \deg(p, \phi_n, D).$$

Justification of the definition. Let d equal the distance between p and the compact set $\phi\ (\partial D)$.

Choose N so large that if $n \geqq N$, then $|\phi_n - \phi| \leqq \frac{1}{2}d$ ($|\ \ |$ stands here for the uniform norm = convergence in the C^0 sense).

Since p does not belong to any ball of radius $\frac{1}{2}d$ and center at $\phi(x)$, $x \in \partial D$, it follows that p is not a convex combination of the form $t\phi_n(x) + (1 - t)\phi_m(x)$, $0 \leqq t \leqq 1$, $x \in \partial D$, n, $m \geqq N$, because $\phi_n(x)$ and $\phi_m(x)$ belong to one such ball.

But then we may fix $n, m \geqq N$ and apply Corollary 3.13 to the family:

$$t\phi_n + (1 - t)\phi_m, \qquad 0 \leqq t \leqq 1.$$

Hence $\deg(p, \phi_n, D) = \deg(p, \phi_m, D)$.
In other words, the limit $\lim_n \deg(p, \phi_n, D)$ does exist and 3.14 is legitimate.

We may reformulate the definition as follows.

3.15. Given $p \notin \phi\ (\partial D)$ there exists a C^0 neighborhood U of ϕ such that for every ψ of class C^1 belonging to U, the degree $\deg(p, \psi, D)$ is the same. Henceforth we define

$$\deg(p, \phi, D) = \deg(p, \psi, D), \qquad \psi \in U.$$

Remark: The correct way to think about the degree is that *it is defined for every continuous function* ϕ and that Definitions II.1 and II.4 in B are only methods of computing it in the special case ϕ C^1.

Our next aim is to extend to the continuous case the statements that we have obtained for the C^1 or C^2 cases. This is done in the following summary theorem.

3.16. Theorem: To every continuous map $\phi : \bar{D} \to R^n$ and every $p \notin \phi\ (\partial D)$ there is associated an integer $\deg(p, \phi, D)$ with the properties:

1. *Invariance under homotopy*. The integer deg (p, ϕ, D) depends only on the homotopy class of ϕ in the following sense: if ϕ_t is a family of mappings $\phi_t \in C$ depending continuously in the uniform topology on the parameter t, $0 \leqq t \leqq 1$, and such that $p \notin \phi_t(\partial D)$ for every t, then

$$\deg(p, \phi_0, D) = \deg(p, \phi_1, D).$$

2. *Dependence only on the boundary values*. As a consequence of (1), we have: if $\phi|_{\partial D} = \psi|_{\partial D}$ and $p \notin \phi(\partial D) = \psi(\partial D)$, then

$$\deg(p, \phi, D) = \deg(p, \psi, D).$$

3. *Continuity*. The function deg (p, ϕ, D) is continuous in ϕ in the uniform topology in the following sense: given ϕ and $p \notin \phi(\partial D)$, there exists a uniform neighborhood U of ϕ such that if $\psi \in U$, then $p \notin \psi(\partial D)$ and

$$\deg(p, \phi, D) = \deg(p, \psi, D).$$

4. If $p \notin \phi(\bar{D})$, then deg $(p, \phi, D) = 0$. If p and q belong to the same component of $R^n - \phi(\partial D)$, then deg $(p, \phi, D) =$ deg (q, ϕ, D).

5. *Decomposition of the domain*. If $D = \bigcup_i D_i$ where each D_i is open, the family $\{D_i\}$ is disjoint, and $\partial D_i \subset \partial D$, then for every $p \notin \phi(\partial D)$:

$$\deg(p, \phi, D) = \sum_i \deg(p, \phi, D_i).$$

6. *The excision property*. If $p \notin \phi(\partial D)$, $K \subset \bar{D}$, K is closed and $p \notin \phi(K)$, then

$$\deg(p, \phi, D) = \deg(p, \phi, D - K).$$

7. *Cartesian products*. If $D \subset R^n$, $D' \subset R^m$ and $\phi : D \to R^n$, $\psi : D' \to R^m$, then

$$\deg((p, q), (\phi, \psi), D \times D') = \deg(p, \phi, D) \deg(q, \psi, D)$$

whenever each term makes sense.

Proof of 3. The proof follows obviously from Definition 3.15.

1 follows from 3. In fact, the function deg (p, ϕ_t, D) is continuous in t. Since the range (the integers) is discrete and $[0, 1]$ is connected, it must be constant.

2 follows from 1. Consider the family $t\phi + (1 - t)\psi$.

Proof of 4. We remarked after Definition II.2 that 4 holds for C^1 mappings. The general case follows immediately by 3 and approximation.

Proof of 5. We assumed that $\bar{D}$ is compact. This can happen only if D_i is a *finite family*. But then we can select a single C^1 mapping ψ such that ψ and each restriction $\psi_i = \psi|_{D_i}$ belong to the C^0 neighborhood corresponding to ϕ and each $\phi|_{D_i}$, respectively, according to Definition 3.15. The statement is thus reduced to the C^1 case, and now it is enough to consider the case $p \notin \psi(Z)$. But then our result follows trivially from the associative law of addition:

$$\deg(p, \psi, D) = \sum_{\psi(x)=p} \operatorname{Sign} J_\psi(x)$$
$$= \sum_i \Bigg(\sum_{\substack{x \in D_i \\ \psi(x)=p}} \operatorname{Sign} J_\psi(x) \Bigg) = \sum_i \deg(p, \psi_i, D_i).$$

Proof of 6. Here again it is easy to see how to reduce the continuous case to the C^1 case. If ϕ is supposed to be C^1, Definition II.1 gives

$$\deg(p, \phi, D) = \sum_{\phi(x)=p} \operatorname{Sign} J_\phi(x)$$

and under the assumption $p \notin \phi(K)$, it follows that

$$\deg(p, \phi, D) = \sum_{\substack{\phi(x)=p \\ x \in D-K}} \operatorname{Sign} J_\phi(x) = \deg(p, \phi, D - K).$$

Proof of 7. Once again the reduction to the C^1 case is immediate and the result follows from the remark that $J_{(\phi,\psi)} = J_\phi \cdot J_\psi$.
Q.E.D.

We now generalize 1 and 2.

3.17. Corollary: If ϕ and ψ have homotopic restrictions to ∂D, i.e., if there exists a family θ_t, $0 \leqq t \leqq 1$ of mappings $\theta_t : \partial D \to R^n$ such that $p \notin \theta_t(\partial D)$ for every t and $\phi|_{\partial D} = \theta_0$, $\psi|_{\partial D} = \theta_1$, then

$$\deg(p, \phi, D) = \deg(p, \psi, D).$$

Proof: Consider the cylinder $L = \bar{D} \times [0, 1]$. The homotopy θ_t may be considered as a continuous function θ from $\partial D \times [0, 1]$ into R^n by defining $\theta(t, x) = \theta_t(x)$. Now extend θ to a mapping T defined on all of L and call $\tilde{\phi}$, $\tilde{\psi}$ the mappings $\tilde{\phi}(x) = T(0, x)$, $\tilde{\psi}(x) = T(1, x)$, $x \in \bar{D}$. Clearly by (3.16; 1) we have

$$\deg(p, \tilde{\psi}, D) = \deg(p, \tilde{\phi}, D).$$

But since $\tilde{\phi}|_{\partial D} = \phi|_{\partial D}$ and $\tilde{\psi}|_{\partial D} = \psi|_{\partial D}$, by (3.16; 2) it also follows that $\deg(p, \phi, D) = \deg(p, \psi, D)$ as desired.

An easy consequence of 3.16 (and one which could have been obtained earlier, if desired) is the Brouwer fixed point theorem and its equivalent "no-retraction" theorem.

3.18. Corollary: ("No-retraction" Theorem.) Let $B \subset R^n$ be the open unit ball. There is no continuous mapping $\phi : \bar{B} \to \partial B$ such that the restriction $\phi|_{\partial B}$ is the identity.

Proof: Under the conditions stated, ϕ should satisfy

$$\deg(p, \phi, B) = \deg(p, id, B)$$

for every properly situated p (by 3.16; 2). The second member is 1 at $p = 0$ (actually at any point in B). This implies, according to (3.16; 4), that $0 \in \phi(B)$. But this contradicts $\phi(B) \subset \partial B$, and we are done.

Q.E.D.

3.19. Theorem: (Brouwer fixed point theorem.) Let B be any open ball in R^n. If $\phi : \bar{B} \to \bar{B}$ is continuous, there exists $x \in \bar{B}$ such that $\phi(x) = x$.

Proof: We may suppose that $\bar{B}$ is the unit ball with center at the origin. If ϕ has no fixed points, then for each $x \in Bx$ and $\phi(x)$ define a straight line. Define $\psi(x)$ as the only point of the form $\lambda x + (1 - \lambda)\phi(x)$, $\lambda \geqq 1$ having norm 1. ψ is continuous, $\psi : \bar{B} \to \partial B$ and $\psi|_{\partial B} = id$. But we have just seen that such a mapping cannot exist.

Q.E.D.

F. The Multiplicative Property and Consequences

As we have seen, if ϕ is a continuous mapping $\phi : \bar{D} \to R^n$, then $\deg(p, \phi, D)$ is constant on every component of $R^n - \phi(\partial D)$ (see (3.16; 4)). This allows us to introduce the notation

$$\deg(A, \phi, D)$$

where A is any non-void set contained in a single component of $R^n - \phi(\partial D)$. The definition is:

$$\deg(A, \phi, D) = \deg(p, \phi, D), \quad p \in A.$$

However, we can point out one distinguished component, namely the unbounded one. Certainly, $\phi(\partial D)$ being compact, there is at least one unbounded component of $R^n - \phi(\partial D)$. But this component contains the exterior of any ball containing $\phi(\partial D)$. Therefore it is unique. Call it Δ_∞. Since $\phi(\bar{D})$ is compact, there exist points in Δ_∞ not in $\phi(\bar{D})$. For any such point p, $\deg(p, \phi, D) = 0$, by (3.16;4). Thus $\deg(\Delta_\infty, \phi, D) = 0$.

3.20. Theorem: (Multiplication Property.) Let $\phi : \bar{D} \to R^n$, $\psi : R^n \to R^n$ be two continuous mappings and Δ_i the bounded components of $R^n - \phi(\partial D)$. Suppose that $p \notin \psi \circ \phi(\partial D)$. Then

$$\deg(p, \psi \circ \phi, D) = \sum_{\Delta_i} \deg(p, \psi, \Delta_i) \deg(\Delta_i, \phi, D).$$

Proof: We place ourselves in the pleasant situation of "counting zeros" by assuming that ϕ and ψ are C^1 mappings and that p is not the image of a critical point. In that case

$$\begin{aligned}
\deg(p, \psi \circ \phi, D) &= \sum_{y;\, \psi(\phi(y)) = p} \operatorname{Sign} J_{\psi \circ \phi}(y) \\
&= \sum_{y;\, \psi(\phi(y)) = p} \operatorname{Sign}(J_\psi(\phi(y))) \operatorname{Sign} J_\phi(y) \\
&= \sum_{\substack{y, z;\, \phi(y) = z \\ \psi(z) = p}} \operatorname{Sign} J_\psi(z) \operatorname{Sign} J_\phi(y) \\
&= \sum_{\substack{z \in R^n - \phi(\partial D) \\ \psi(z) = p}} \deg(z, \phi, D) \operatorname{Sign} J_\psi(z).
\end{aligned}$$

But $R^n - \phi(\partial D)$ is the union of the disjoint sets Δ_i, so that

$$\begin{aligned}
\deg(p, \psi \circ \phi, D) &= \sum_{\Delta_i} \left(\deg(\Delta_i, \phi, D) \sum_{\substack{z \in \Delta_i \\ \psi(z) = p}} \operatorname{Sign} J_\psi(z) \right) \\
&= \sum_{\Delta_i} \deg(\Delta_i, \phi, D) \deg(p, \psi, \Delta_i),
\end{aligned}$$

and the proof is done.

As an illustration of the power of this theorem it is possible we give, following Leray, an immediate proof of the Jordan separation theorem (cf. Jean Leray, Proc. Int. Congress, 1950, vol. II, pp. 202–208).

3.21. Theorem: (Jordan) Let K and L be homeomorphic compact sets in R^n. Then $R^n - K$ and $R^n - L$ have the same number of components.

Proof: (Leray) Suppose that $h : K \to L$ is a homeomorphism and that ϕ and ψ are extensions respectively of h and h^{-1} to all of R^n. Let Δ_i be the components of $R^n - K$ and D_i those of $R^n - L$. Now the mapping $\psi \circ \phi : R^n \to R^n$ is the identity on K. Consider Δ_j. The restriction $\psi \circ \phi|_{\partial \Delta_i}$ is also the identity since $\partial \Delta_i \subset K$. Therefore if p is properly located

$$\deg(p, \psi \circ \phi, \Delta_i) = \deg(p, id, \Delta_i)$$

by (3.16; 2).

But for every i, the degree $\deg(\Delta_i, id, \Delta_j)$ is equal to δ_{ij} (Kronecker's δ), and then:

$$\deg(\Delta_j, \psi\circ\phi, \Delta_i) = \delta_{i,j}.$$

Using the multiplicative property we shall obtain

$$\delta_{i,j} = \sum_h \deg(\Delta_j, \psi, D_h) \deg(D_h, \phi, \Delta_i).$$

This is not immediate, because the sets D_h are not the components of $R^n - \phi(\partial\Delta_i)$, but subsets of them.

Fix j and compare the sets:

$$R^n - L = \bigcup D_i$$

$$R^n - \phi(\partial\Delta_j) = \bigcup G_i,$$

where G_i are the components of $R^n - \phi(\partial D_j)$. The sets of the family $\{G_i\}$ are disjoint.

Since L and $\phi(\partial\Delta_j)$ are compact there is only one unbounded component in each case. Call these Δ_∞ and G_∞. Since $\phi(\partial\Delta_j) \subset L$, or

$$R^n - L \subset R^n - \phi(\partial\Delta_j).$$

It follows by connectedness arguments that the family $\{D_i\}$ splits into several subfamilies $\{D_j^i\}$ such that

$$\begin{array}{ccc} U^1 = D_1^1 \cup \cdots \cup D_n^1 \cup \cdots \subset G_1 \\ U^2 = D_1^2 \cup \cdots \cup D_n^2 \cup \cdots \subset G_2 \\ \vdots \quad \vdots \qquad\qquad \vdots \end{array}$$

and furthermore, that D_∞ is necessarily contained in G_∞. Consequently

(1) $$\deg(D_h^i, \phi, \Delta_j) = \deg(G_i, \phi, \Delta_j).$$

Let $M_i = \bar{G}_i - U^i$. It is easy to see that $M_i \subset L$ for every i. But now if $p \in \Delta_i$ and $\psi(z) = p$, z cannot belong to L, because $\psi(L) = h(L) = K$, and, a posteriori, z cannot belong to any M_i either. This implies that

$$\deg(p, \psi, G_i) = \deg(p, \psi, G_i - M_i)$$

as was seen in (3.16; 6).

But then, since $G_i - M_i = \bigcup_j D_j^i$, we conclude (3.16; 5) that

(2) $$\deg(p, \psi, G_i) = \deg(p, \psi, \bigcup_j D_j^i) = \sum_j \deg(p, \psi, D_j^i).$$

Now we recall that the multiplicative property means exactly

$$\deg(p, \psi\circ\phi, \Delta_j) = \sum_{h\neq\infty} \deg(p, \psi, G_h)\deg(G_h, \phi\,\Delta_j),$$

and, by (1) and (2), it follows that

$$\begin{aligned}\deg(p, \psi\circ\phi, \Delta_j) &= \sum_i \Big(\sum_h \deg(p, \psi, D_h^i)\Big)\deg(D_h^i, \phi, \Delta_j)\\ &= \sum_{D_i\neq D_\infty} \deg(p, \psi, D_i)\deg(D_i, \phi, \Delta_j).\end{aligned}$$

Since p was any point in Δ_i, we conclude

$$(3)\qquad \delta_{i,j} = \deg(\Delta_i, \psi\circ\phi, \Delta_j) = \sum_{D\neq D_\infty} \deg(\Delta_i, \psi, D)\deg(D, \phi, \Delta_j).$$

By the symmetry of the argument it is also true that

$$(4)\qquad \delta_{i,j} = \sum_{\Delta\neq\Delta_\infty} \deg(D_i, \phi, \Delta)\deg(\Delta, \psi, D_j).$$

Suppose that both families $\{\Delta_i\}$ and $\{D_i\}$ are finite. Then they define two matrices

$$A = (\deg(D_i, \phi, \Delta_j))$$

$$B = (\deg(\Delta_i, \psi, D_j))$$

and (3) and (4) imply

$$(5)\qquad \begin{aligned} AB &= I_n\\ BA &= I_m\end{aligned}$$

if we assume A to be $n \times m$ and B to be $m \times n$, where n = number of D_i's, m = number of Δ_i's. But now it is an easy exercise to show that the equalities (5) imply $n = m$, and this is what we were looking for.

The case $\{D_i\}$ finite, $\{\Delta_i\}$ infinite is easily seen to be impossible by the equalities (3) and (4). If both families are infinite, then being disjoint families of open sets, are both denumerably infinite.

Q.E.D.

We can now draw some topological consequences.

3.22. Corollary: (Domain Invariance) Let U be an open set in R^n, $\phi: U \to R^n$ a continuous and one-to-one mapping. Then ϕ is an open mapping.

Proof: Pick a point $p \in U$ and let $\bar{D}$ be the closure of an open ball D containing p and contained in U.

Since $\bar{D}$ is compact, $\phi : \bar{D} \to \phi(\bar{D})$ is a homeomorphism and we can apply Jordan's theorem which implies that $R^n - \phi(\bar{D})$ has one component and that $R^n - \phi(\partial D)$ has two components.

Take a point q in the bounded component Δ of $R^n - \phi(\partial D)$. If $q \in R^n - \phi(\bar{D})$, it follows from the connectedness of $R^n - \phi(\bar{D})$ that q can be joined by a continuous curve with every other point in $R^n - \phi(\bar{D})$. But this set certainly intersects the unbounded component of $R^n - \phi(\partial D)$ which is a contradiction. We conclude that $\Delta \cap (R^n - \phi(\bar{D})) = \varnothing$ or $\Delta \subset \phi(\bar{D})$. As we know that $\phi(p) \notin \phi(\partial D)$, because ϕ is $1-1$, $\phi(p)$ must belong to Δ. Hence ϕ is open.

Q.E.D.

G. Borsuk's Theorem

We shall prove the following

3.23. Theorem: Let D be a symmetric bounded open set in R^n containing the origin and $\phi : D \to R^n$ an *odd* mapping ($\phi(x) = -\phi(-x)$, for all $x \in D$) such that $0 \notin \phi(\partial D)$. Then $\deg(0, \phi, D)$ is an odd number (in particular different from zero).

The proof follows from a sequence of lemmas.

3.24. Lemma: Let $K \subset R^n$ be a compact set, $\phi : K \to R^{n+1}$ a continuous mapping such that $0 \notin \phi(K)$. Then ϕ can be extended to a continuous never vanishing mapping defined on a cube $C \supseteq K$.

Proof: For $\varepsilon > 0$ choose ψ to be C^1 and such that $|\phi(x) - \psi(x)| < \varepsilon$ for $x \in K$. $\psi(D)$ has measure zero for every $D \subset R^n$ by Sard's lemma, and so it is possible to pick a point y_0 such that $x \to \psi(x) + y_0$ never assumes the value 0. Suppose then that ψ itself is never vanishing.

Let $c = \inf|\phi(x)|$, $x \in K$ and choose a continuous function defined for $t > 0$ with values in R such that

$$\eta(t) = 1 \quad \text{if} \quad t \geqq \frac{c}{2} \qquad\qquad \eta(t) = \frac{2t}{c} \quad \text{if} \quad t \leqq \frac{c}{2}.$$

If we define Φ as the mapping

$$\Phi(x) = \frac{\psi(x)}{\eta(|\psi(x)|)},$$

then $|\Phi(x)| \geqq c/2$ for all x and $|\Phi(x) - \phi(x)| < c$ on K.

Suppose that ε has been chosen so that $\varepsilon < c/2$.

By the Tietze extension theorem there exists a function $\delta : C \to R^n$ such that $\delta(x) = \Phi(x) - \phi(x)$ if $x \in K$, and $|\delta(x)| \leqq \varepsilon$ for $x \in C$. Define $\phi_1(x) = \Phi(x) - \delta(x)$, $x \in C$.

Then

$$\phi_1(x) = \phi(x) \quad \text{if} \quad x \in K,$$

$$|\phi_1(x)| = |\Phi(x) - \delta(x)| \geqq |\Phi(x)| - |\delta(x)| \geqq \frac{c}{2} - \varepsilon > 0$$

and ϕ_1 is a solution of our problem.

3.25. Lemma: Let $D \subset R^n$ be a symmetric open bounded set such that $0 \notin D$. Let ϕ be a mapping of ∂D in R^m, $m > n$, which is odd and non-vanishing. Then ϕ can be extended to $\bar{D}$ to be odd and non-vanishing.

Proof: We shall use the induction on the dimension of R^n. For $n = 1$, D looks like:

$$\varepsilon$$

$$— — \leftrightarrow — \leftrightarrow — \leftrightarrow — — — \leftrightarrow — \leftrightarrow — \leftrightarrow — —$$

By the previous lemma we can extend the function $\psi = \phi$ restricted to $[\varepsilon, \infty] \cap \partial D$, to a non-vanishing function $\tilde{\psi}$ defined on some interval $[\varepsilon, N]$. By symmetry, we may define a function $\tilde{\phi}$ extending ϕ and never vanishing.

Suppose now that the lemma is established for $n_1 < n$. Let $x \in R^n$, $\hat{x} \in R^{n-1}$ (suppose furthermore that R^{n-1} has been identified with the hyperplane $x_1 = 0$ in R^n). Considering $R^{n-1} \cap D, \phi$ can be extended to $\partial D \cup (R^{n-1} \cap D)$ to be odd and non-vanishing (this is our inductive step): call the extension again ϕ.

Now split R^n into R^{n-1}, R^n_+, R^n_- where $x_1 = 0$, $x_1 > 0$, $x_1 < 0$ respectively, and let $D_+ = D \cap R^n_+$, $D_- = D \cap R^n_-$. By the previous lemma ϕ has a further extension to $\partial D \cup (R^{n-1} \cap D) \cup D_+$, continuous and non-vanishing. Now, by symmetry, the final extension can be defined.

Q.E.D.

3.26. Lemma: Let $D \subset R^n$ be a bounded open symmetric set such that $0 \notin \bar{D}$, $\phi : \partial D \to R^n$, a continuous odd and never vanishing mapping. Then ϕ can be extended to $\bar{D}$ to be continuous and odd, and furthermore, non-vanishing on $\bar{D} \cap R^{n-1}$ (again the identification $R^{n-1} \subset R^n$).

It follows from the previous lemma applied to ϕ retsricted to $\partial (D \cap R^{n-1}) = \partial D \cap R^{n-1}$, that we can obtain a never-vanishing extension to $\bar{D} \cap R^{n-1}$. Such an extension can be extended at once to the desired map on $\bar{D}$ by symmetry.

Q.E.D.

3.27. Lemma: If $D \subset R^n$ is a bounded open symmetric set and $0 \notin \bar{D}$, for every $\phi : \partial D \to R^n$ continuous and odd such that $0 \notin \phi(\partial D)$, $\deg(0, \phi, D)$ is an even integer.

Proof: Extend ϕ to $\bar{D}$ so as to be a continuous odd mapping never vanishing on $\bar{D} \cap R^{n-1}$. The lemma above assures the existence of such an extension. Call the extended mapping also ϕ. Approximate ϕ by a mapping ψ of class C^1 and *odd* (replace, if necessary, an approximating ψ by its odd part $\frac{1}{2}[\psi(x) - \psi(-x)]$). If ψ is close enough to ϕ, it follows that

$$0 \notin \psi(\partial D)$$

$$0 \notin \psi(\bar{D} \cap R^{n-1})$$

$$\deg(0, \psi, D) = \deg(0, \phi, D).$$

We want to compute $\deg(0, \psi, D)$. Consider the sets $D^+ = R^n_+ \cap D$, $D^- = R^n_- \cap D$ (where $R^n_+ = \{x | x_1 > 0\}$, $R^n_- = \{x | x_1 < 0\}$).

By construction ψ never vanishes on $\bar{D} \cap R^{n-1}$, so we can avoid this set and obtain:

$$\text{(1)} \qquad \begin{aligned} \deg(0, \psi, D) &= \deg(0, \psi, D^+ \cup D^-) \\ &= \deg(0, \psi, D^+) + \deg(0, \psi, D^-). \end{aligned}$$

Choose p close to 0 and such that p is not the image under ψ of a critical point of ψ. Observe now that since ψ is *odd*, each partial derivative $\partial\psi/\partial x_i$ is even. But then J_ψ is also an even mapping. This implies in partiçular that $-p$ is not the image of a critical point either. Compute

$$\deg(0, \psi, D_+) = \sum_{\substack{\psi(y)=p \\ y \in D_+}} \text{Sign } J_\phi(y),$$

$$\deg(0, \psi, D_-) = \sum_{\substack{\psi(z)=-p \\ z \in D_-}} \text{Sign } J_\phi(z).$$

Since ψ is odd, the set $\{z | \psi(z) = p, z \in D_-\}$ can be obtained by taking the opposite of the elements in $\{y | \psi(y) = -p\}$. But $J_\psi(z) = J_\psi(-z)$ and we conclude that

$$\deg(0, \psi, D_+) = \deg(0, \psi, D_-).$$

Then (1) implies that $\deg(0, \phi, D) = \deg(0, \psi, D)$ is an *even number*.
Q.E.D.

Now we are ready to prove Borsuk's theorem.

Consider a small open ball U with center at 0, and a mapping $f: \bar{D} \to R^n$ such that

function follows from the observation that if g is an
) and (c) (such an extension certainly exists) then
satisfies (a), (b) and (c).

$= \phi|_{\partial D}$ implies

$$\deg(0, f, D) = \deg(0, \phi, D)$$

; 2).

, it is clear that $f \neq 0$ on ∂U and then

$$(0, f, D) = \deg(0, f, U \cup (D - \bar{U}))$$

$$= \deg(0, f, U) + \deg(0, f, D - \bar{U})$$

$$= 1 + \deg(0, f, D - \bar{U}).$$

n is known to be even by the last lemma. This proves that

$$\deg(0, \phi, D) = \deg(0, f, D)$$

is odd.

Q.E.D.

We now draw some consequences from Borsuk's theorem.

3.28. Corollary: Let D be as in Borsuk's theorem and $\psi: \partial D \to R^n$ a continuous mapping. Then there exists no homotopy ψ_t of ψ into a constant mapping such that $\psi_t(x) \neq 0$ for all t, $x \in \partial D$.

Proof: First extend ψ to some ϕ defined on $\bar{D}$. Replacing ϕ by $\frac{1}{2}(\phi(x) - \phi(-x))$ we may suppose that ϕ itself is odd. But then Borsuk's theorem implies that $\deg(0, \phi, D) \neq 0$ and the impossibility of the existence of the homotopy described is then apparent.

Q.E.D.

3.29. Corollary: Let D be as above. If $\psi: \partial D \to R^n$ is an odd continuous mapping whose image is contained in a subspace $E \neq R^n$, then ψ assumes the value 0 at some point of ∂D.

Proof: Extend ψ to a continuous odd mapping $\phi: \bar{D} \to E$. If $0 \notin \psi(\partial D)$, then by Borsuk's theorem, $\deg(0, \phi, D)$, being odd, is different from zero.

But this implies that $\deg(p, \phi, D)$ differs from zero on the component Δ of $R^n - \psi(\partial D)$ containing 0. Now (3.16; 4) implies that $\phi(\bar{D})$ contains such a component, and, a posteriori, E does also. But this is impossible, since Δ is open, non-void, and $E \neq R^n$.

Q.E.D.

3.30. Corollary: Let D be as above, and let ψ any continuous mapping $\psi: \partial D \to R^n$ whose image is contained in a subspace $E \neq R_n$. Then there exists $p \in \partial D$ such that $\psi(p) = \psi(-p)$.

Proof: Apply the corollary above to the mapping $\frac{1}{2}(\psi(x) - \psi(-x))$.
Q.E.D.

3.31. Corollary: Let D be as above, and $\phi: \bar{D} \to R^n$ a continuous mapping never vanishing on ∂D, such that for every $x \in \partial D$,

$$(1) \qquad \alpha\phi(x) \neq (1-\alpha)\phi(-x)$$

for all α, $\frac{1}{2} \leqq \alpha \leqq 1$.

Then $\phi(D)$ contains a neighborhood of the origin.

Proof: Observe that the conclusion follows from the statement

$$\deg(0, \phi, D) \neq 0.$$

This property is an immediate consequence of the fact that if ψ is the mapping $\psi(x) = \frac{1}{2}(\phi(x) - \phi(-x))$, then by Borsuk's theorem, $\deg(0, \psi, D) \neq 0$. Under condition (1), the family

$$t\phi(x) - (1-t)\phi(-x), \quad \frac{1}{2} \leqq t \leqq 1,$$

is a homotopy between ϕ and ψ, which implies

$$\deg(0, \phi, D) = \deg(0, \psi, D).$$

Q.E.D.

Degree Theory—General Case

H. Preliminaries: Degree Theory in an Arbitrary Finite Dimensional Vector Space

Suppose E is a real vector space of dimension n. By choosing a basis in E we can identify E with R^n. This should allow us to define deg (p, ϕ, D) as it was done for R^n and of course the only important thing is to see what happens after a change of basis. The answer is that the degree is basis–independent. More precisely, given a basis $B = \{b_1, \ldots, b_n\}$, we shall for the moment denote by deg $B(p, \phi, D)$ the degree computed with respect to B; then we have

3.32. Proposition: For every pair of bases B, P

$$\deg_B (p, \phi, D) = \deg_P (p, \phi, D),$$

whenever the expressions make sense.

Proof: It suffices to prove this for C^1 mappings. But then we only need to know what happens to the sign of the Jacobian of a mapping when the basis is changed. This is easily seen to be invariant, whence the result.
Q.E.D.

I. Preliminaries: Restriction to a Subspace

Suppose that $D \subset R^n$ is an open and bounded set, and that $R^m \subseteq R^n$, where the inclusion is made by identifying R^m with the subspace of R^n whose points are the x such that $x_{m+1} = x_{m+2} = \cdots = x_n = 0$.

3.33. Proposition: If $\phi : \overline{D} \to R^m$ is continuous and $\psi : \overline{D} \to R^n$ is the mapping $\psi = id + \phi$, for every $p \in R^m$ not belonging to $\psi(\partial D)$:

$$\deg (p, \psi, D) = \deg (p, \psi|_{R^m \cap \overline{D}}, D \cap R^m).$$

Proof: Let us begin by noting that $\psi(R^m \cap \overline{D}) \subset R^m$ as can be verified easily; thus the expression deg $(p, \psi|_{R^m \cap \overline{D}}, R^m \cap D)$ makes sense.

Suppose that ϕ is C^1. Bv definition it suffices to prove the statement for this case, and under the assumption that p is not the image of a critical point of ψ. As the degree is then computed by counting zeros, it is necessary to look for the points y in $\psi^{-1}(p)$. If $\psi(y) = y + \phi(y) = p$, then $y = p - \phi(y) \in R^m$. Hence $\psi^{-1}(p) \subset R^m \cap D$.

This implies that the points to be counted for $\psi \colon \bar{D} \to R^n$ and for $F = \psi|_{R^m \cap \bar{D}}$, $F \colon R^m \cap \bar{D} \to R^m$ *are the same*, and the only possible difference in degree lies in the signs assigned to them.

Our proposition will follow from the fact that at each such point y, we have

$$\text{Sign } J_F(y) = \text{Sign } J_\psi(y). \tag{1}$$

To prove (1), first observe that the Jacobian matrix of ψ has the form

$$\psi' = \left[\begin{array}{c|c} 1 + \dfrac{\partial \phi_i}{\partial x_j} & 0 \\ \hline 0 & I \end{array}\right] \quad \Big] \, m$$

(j runs from 1 to m).

This implies immediately that $J_\psi(x) = J_F(x)$ for every $x \in R^m \cap D$, which clearly implies (1) above. Hence our proposition has been proved.

Q.E.D.

J. Degree of Finite Dimensional Perturbations of the Identity

Let X be a real Hausdorff locally convex T.L.S., and $D \subset X$ an open subset of X such that $E \cap D$ is bounded for every finite dimensional subspace E of X (in that case we say that D is "finitely bounded"). This is the most general case we shall consider.

We now give some definitions.

3.34. Definition: If T is a topological space and $\phi : T \to X$ is a continuous mapping, we shall say that ϕ is *finite dimensional* if $\phi(T)$ is contained in some finite dimensional subspace of X. If T is also a subset of X, we define a *finite dimensional perturbation* of the identity to be a mapping $\psi : T \to X$ of the form $\psi = 1 + \phi$ where 1 is the identity $1 : T \to X$ and ϕ is finite dimensional. $\phi = \psi - 1$ is called the *perturbation* of ψ.

Our aim is to define the degree $\deg(p, \psi, D)$ for every finite dimensional perturbation of the identity $\psi = 1 + \phi$ (defined on $T = \bar{D}$, D as above). Let

$p \in X$ be a point not in $\psi(\partial D)$ and choose a finite dimensional subspace $E \subset X$ such that

$$\text{(a)} \qquad p \in E,$$

$$\phi(\bar{D}) \subset E.$$

Letting f denote the restriction $f\colon \bar{D} \cap E \xrightarrow{\psi} E$, we have

3.35. Definition:

$$\deg(p, \psi, D) = \deg(p, f, D \cap E),$$

where the second member is computed in E according to the theory for finite dimensional spaces.

We must justify 3.35. Suppose F is a subspace of X satisfying properties (a) above and that $F \subset E$. Then proposition 3.32 applies and we conclude

$$\deg(p, f|_{F \cap D}, D \cap F) = \deg(p, f, D \cap E).$$

If F satisfies (a) but F and E are not nested, we reduce to that case by considering $E \subset E + F$ and $F \subset E + F$ separately. Thus 3.35 is justified.

3.36. Remark: We know that in the finite dimensional case the degree $\deg(p, \phi, D)$ depends only on the restriction of ϕ to ∂D (see 3.16; 2).

Let us suppose that we have a finite dimensional mapping $\tilde{\phi}$ defined only on ∂D, $\tilde{\phi}\colon \partial D \to X$. The degree of all finite dimensional perturbations of the identity $1 + \phi$ by means of a finite dimensional extension ϕ of $\tilde{\phi}$ defined on all of D will be the same as follows from Definition 3.35 and the finite dimensional theory. But we cannot assure the *existence* of such extensions unless we assume X to have additional properties (for instance, to be normal). Nevertheless a notion of degree of $1 + \tilde{\phi}$ may be defined. Suppose $p \notin \phi(\partial D)$. Choose a finite dimensional subspace E containing both p and $\phi(\partial D)$. E is finite dimensional and so there are extensions ϕ of $\tilde{\phi}|_{\partial D \cap E}$ to all of $D \cap E$. Thus it is possible to define $\deg(p, \tilde{\phi} + 1, D)$ by

$$\deg(p, \tilde{\phi} + 1, D) = \deg(p, \tilde{\phi} + 1, D),$$

where the second member is computed in E.

Hence whenever we have a finite dimensional mapping $\tilde{\phi}\colon \partial D \to X$ *we can define the degree* $\deg(p, \tilde{\phi} + 1, D)$, *and this coincides with the degree of every finite dimensional perturbation of* 1 *by means of an extension of* $\tilde{\phi}$ *to all of* $\bar{D}$.

K. Properties

We have shown that the definition of degree can be generalized to obtain a notion of degree for finite dimensional perturbations of the identity with respect to domains "finitely bounded" in an arbitrary locally convex T.L.S. Now we shall list the properties of degree that remain valid in this situation.

From the Definition 3.35 and 3.16; 4, 5, 6 and 7 we obtain:

3.37. Proposition: For every finite dimensional perturbation of the identity $\psi = 1 + \phi : \bar{D} \to X$, and every $p \notin \psi(\partial D)$, the following results hold:

1. If $p \notin \psi(D)$, then $\deg(p, \psi, D) = 0$.
2. If p and q belong to the same component of $X - \psi(\partial D)$, then
$$\deg(p, \psi, D) = \deg(q, \psi, D).$$
3. If $D = \bigcup_i D_i$, where the family $\{D_i\}$ is disjoint and $\partial D_i \subset \partial D$, then
$$\deg(p, \psi, D) = \sum_i \deg(p, \psi, D_i).$$
4. If $K \subset \bar{D}$, K is closed, $p \notin \psi(K)$, then
$$\deg(p, \psi, D) = \deg(p, \psi, D - K).$$
5. If $f: \bar{D}_1 \to X_1$ is a mapping satisfying the same conditions as ϕ, then
$$\deg((p \times q), 1 + (\phi \times f), D \times D_1) = \deg(p, 1 + \phi, D) \deg(q, 1 + f, D_1)$$
whenever these expressions make sense.

L. Limits

The family of finite dimensional mappings is closed under addition and product by scalars. Hence to proceed we consider limits of such mappings. From this point of view the important thing is that the *compact mappings* (definition below) are such limits and so we will be able to define degrees of compact perturbations of the identity.

Let us recall and introduce some notations: D is an open set in X such that $D \cap E$ is bounded for every finite dimensional subspace $E \subset X$. $C(\bar{D})$ will denote the (linear) space of all continuous mappings $\phi : \bar{D} \to X$; likewise, $C(\partial D)$ will be the space of continuous mappings $\phi : \partial D \to X$. There exists a natural mapping $Q : C(\bar{D}) \to C(\partial D)$ defined by restriction $Q(\phi) = \phi|_{\partial D}$.

Similarly, we denote by $\mathscr{F}(\bar{D})$ and $\mathscr{F}(\partial D)$ the subspaces of $C(\bar{D})$ and $C(\partial D)$ whose elements are the finite dimensional mappings. Of course, $Q: \mathscr{F}(\bar{D}) \to \mathscr{F}(\partial D)$.

Let us give $C(\bar{D})$ and $C(\partial D)$ the topologies of the *uniform convergence.* The open sets of, say, $C(\bar{D})$, are those defined by

$$\{\phi \mid \phi(\bar{D}) \subset G\}$$

where G runs over all the open sets of X.

Warning: These topologies are *not* linear space topologies, but merely *group* topologies (i.e., the mapping $(\phi, \psi) \to \phi + \psi$ is continuous, while the mapping $(\lambda, \phi) \to \lambda\phi$, $\lambda \in R, \phi \in C(\bar{D})$, is not necessarily continuous).

By 3.36, for every $\phi \in \mathscr{F}(\partial D)$ and every $p \notin (1 + \phi)(\partial D)$, the degree $\deg(p, 1 + \phi, D)$ is defined, and for every $g \in C(\bar{D})$ such that $Qg = \phi$,

$$\deg(p, 1 + g, D) = \deg(p, 1 + \phi, D).$$

3.38. Lemma: Let $\phi \in C(\partial D)$, p be a point of X and V be a convex symmetric neighborhood of $0 \in X$ such that:

(a) $$(p + V) \cap (1 + \phi)(\partial D) = \varnothing.$$

Then, if $f \in \mathscr{F}(\partial D)$ satisfying $f(x) - \phi(x) \in V$ for every $x \in \partial D$, the degrees

$$\deg(p, 1 + f, D)$$

are defined; moreover, these degrees are equal for all such f.

Proof: Suppose $p = x + f(x), x \in \partial D$. Then $x + \phi(x) = p + (\phi(x) - f(x)) \in p + V$, which contradicts (a). Hence $p \notin (1 + f)(\partial D)$ and the degree is defined.

Suppose that $g \in \mathscr{F}(\partial D)$ also satisfies $g(x) - \phi(x) \in V$ for every $a \in \partial D$, and call $F = 1 + f$, $G = 1 + g$, $\psi = 1 + \phi$.

For every $x \in \partial D$, $F(x)$ and $G(x)$ both belong to $\psi(x) + V$ and this set is convex. Hence any convex combination $(1 - t)G(x) + tF(x)$, $0 \leqq t \leqq 1$ also belongs to $\psi(x) + V$. Using (a) this implies that for every t, $0 \leqq t \leqq 1$, and every $x \in \partial D$,

$$p \neq (1 - t)G(x) + tF(x).$$

Let E be a finite dimensional subspace of X such that

$$p \in E$$

$$G(\partial D) \subset E$$

$$F(\partial D) \subset E.$$

Considering the domain $D \cap E$ and the homotopy $(1 - t) G + tF$ between G and F, we conclude by 3.17 that $\deg(p, G, D \cap E) = \deg(p, F, D \cap E)$, or, according to Definition 3.35, $\deg(p, G, D) = \deg(p, F, D)$ as desired.
Q.E.D.

3.39. Proposition: Let $\phi \in C(\partial D)$ and p a point of X not in the closure of $(1 + \phi)(\partial D)$. Then there exists a neighborhood U of ϕ in $C(\partial D)$ such that the degree

$$\deg(p, 1 + f, D)$$

takes on but a single value for all f belonging to $U \cap \mathscr{F}(\partial D)$.

Proof: Follows from the lemma above.
Q.E.D.

This proposition will permit us to define the degree for perturbations of the identity by limits of finite dimensional mappings.

Call $\mathscr{L}(\partial D)$, $\mathscr{L}(\overline{D})$ the closure of $\mathscr{F}(\partial D)$ (respectively $\mathscr{F}(\overline{D})$) in $C(\partial D)$ (respectively $C(\overline{D})$). Plainly $Q : \mathscr{L}(\overline{D}) \to \mathscr{L}(\partial D)$.

3.40. Definition: Let $\phi = \psi - 1$ be a mapping in $\mathscr{L}(\partial D)$. Let p be a point of X not in the closure of $\psi(\partial D)$. The common value of the degrees $\deg(p, 1 + f, D)$ when $f \in \mathscr{F}(\partial D)$ is near ϕ is defined to be $\deg(p, \psi, D)$.

If $\phi = \psi - 1$ is a mapping defined on all of $\overline{D}$ and such that $Q(\phi) \in \mathscr{L}(\partial D)$, we shall write simply $\deg(p, \psi, D)$ instead of $\deg(p, Q\psi, D)$. In particular, for every $\phi \in \mathscr{L}(\overline{D})$, the degree is defined.

From 3.37 the next proposition follows easily.

3.41. Proposition: If $\phi = \psi - 1 \in \mathscr{L}(\overline{D})$ and p and q do not belong to the closure of $\psi(\partial D)$, then:

1. If $p \notin \overline{\psi(D)}$, then $\deg(p, \psi, D) = 0$.
2. If p and q belong to the same component of $X - \overline{\psi(\partial D)}$, then

$$\deg(p, \psi, D) = \deg(q, \psi, D).$$

3. If $D = \bigcup D_i$, where the family $\{D_i\}$ is disjoint and $\partial D_i \subset \partial D$, then

$$\deg(p, \psi, D) = \sum_i \deg(p, \psi, D_i).$$

4. If $K \subset \overline{D}$, K is closed and $p \notin \overline{\psi(K)}$, then

$$\deg(p, \psi, D) = \deg(p, \psi, D - K).$$

5. If $f : \overline{D}_1 \to X_1$ is a mapping of $\mathscr{L}(\overline{D}_1)$, then

$$\deg((p, p_1), 1 + (\phi, f), D \times D_1) = \deg(p, 1 + \phi, D) \deg(p_1, 1 + f, D_1).$$

Proof: Left to the reader.
(Hint: Approximate and use Proposition 3.16.)
In the same way it is possible to prove the following generalization of 3.33:

3.42. Proposition: Let Y be a closed subspace of X and p be a point of Y not in $\overline{\psi(\partial D)}$, where $\psi - 1 \in \mathscr{L}(\partial D)$. Then

$$\deg(p, \psi, D) = \deg(p, \psi|_{\partial D \cap Y}, D \cap Y).$$

Finally, we have a generalization of Borsuk's theorem:

3.43. Proposition: If $D \subset X$ is an open set which is finitely bounded, symmetric, and contains the origin, then for every odd ψ such that $\psi - 1 \in \mathscr{L}(\partial D)$, and if $0 \notin \overline{\psi(\partial D)}$, then $\deg(0, \psi, D)$ is an odd number.

The proof follows at once from the Definition 3.40 and the Borsuk theorem 3.23.

3.44. Corollary: If D is a domain as in the proposition above, if $\psi - 1$ is odd and belongs to $\mathscr{L}(\bar{D})$, and if $0 \notin \overline{\psi(\partial D)}$, then $\psi(D)$ covers a neighborhood of 0.

M. Compact Perturbations

Here we shall show that the compact mappings are in $\mathscr{L}(\partial D)$ and $\mathscr{L}(\bar{D})$ and obtain some additional properties of the degree of compact perturbations of the identity.

We begin with some purely topological results.

Let X be a locally convex T.L.S., T a topological space. Let $C(T)$ denote the set of continuous mappings $\phi: T \to X$. $C(T)$ has a natural structure as a topological space (see the beginning of section L).

Consider the subspace $\mathscr{F}(T)$ of $C(T)$ whose elements are the mappings ϕ such that $\phi(T)$ is contained in a finite dimensional subspace of X and the subspace $K(T)$ of the mappings ϕ for which the set $\phi(T)$ is pre-compact.

3.45. Proposition:

$$K(T) \subset \overline{\mathscr{F}(T) \cap K(T)}.$$

Proof: Choose an open, convex, symmetric neighborhood V of 0 in X, and let $\phi \in K(T)$. Suppose the points $y_1, \ldots, y_n \in X$ have the property

$$\phi(T) \subset \bigcup_{i=1}^{n} \{y_i + V\}.$$

Letting ϱ be the gauge induced by V,

$$\varrho(x) = \inf_{x \in \lambda V} |\lambda|, \qquad x \in X$$

we define mappings $\mu_i : T \to R$ by

$$\mu_i(x) = \max(0, 1 - \varrho(\phi(x) - y_i)).$$

Each μ_i is *continuous* (since ϱ is continuous).

Since $\phi(T) \subset \bigcup \{y_i + V\}$, for each x there exists at least one $\mu_i(x)$ different from zero. Thus the function $\mu(x) = \sum_i \mu_i(x)$ never vanishes on $\phi(T)$, and we can define

$$\lambda_i(x) = \frac{\mu_i(x)}{\mu(x)}.$$

These mappings satisfy $1 \geqq \lambda_i \geqq 0$, $\sum_i \lambda_i(x) = 1$. Now define

$$\phi_\varepsilon(x) = \sum_{i=1}^{n} \lambda_i(x)\, y_i.$$

This mapping belongs to $\mathscr{F}(T) \cap K(T)$ and

$$\phi(x) - \phi_\varepsilon(x) = \sum_i \lambda_i(x)\,(\phi(x) - y_i).$$

We see that if $\phi(x) - y_i \notin V$, then $\varrho(\phi(x) - y_i) \geqq 1$, and consequently $\mu_i(x) = 0$, which implies $\lambda_i(x) = 0$. This means that $\phi(x) - \phi_\varepsilon(x)$ is a convex combination of elements of V which belongs to V since V is convex. Thus ϕ is a limit point of $\mathscr{F}(T) \cap K(T)$, as desired.

Q.E.D.

Using this proposition we may return to our initial situation $D \subset X$, D an open finitely bounded set.

We shall say that a mapping $\phi \in C(\bar{D})$ (or $\phi \in C(\partial D)$) is *compact* iff $\overline{\phi(D)}$ is a compact set (respectively: $\phi(\partial D)$).

(Not to be confused with the mappings of Definition 1.38.)

3.46. Proposition: Any compact mapping $\phi \in C(\bar{D})$ (respectively $\phi \in C(\partial D)$) belongs to $\mathscr{L}(\bar{D})$ (respectively to $\mathscr{L}(\partial D)$).

Proof: Immediate from 3.45.

Q.E.D.

The perturbations of the identity by elements of $\mathscr{L}(\bar{D})$ do not behave "nicely" topologically and in the preceding it was necessary to consider such artificial sets as the closure of $(1 + \phi)(\partial D)$. The compact ones however look like finite dimensional mappings, and we have

3.47. Proposition: If D is any domain $D \subset X$, and $\phi \in C(\partial D)$ (respectively $\phi \in C(\bar{D})$) is compact, then $\psi = 1 + \phi$ is proper and closed.

Proof: To say that ψ is proper means that the inverse image of a compact set is also compact. Suppose $K \subset K$ is compact, and let $A = \psi^{-1}(K)$. Suppose $\{x_\alpha\}$ is an indexed family in A. Then $\{x_\alpha + \phi(x_\alpha)\}$ being contained in K, has a convergent subfamily $x_\beta + \phi(x_\beta) \to y$. But ϕ being a compact mapping, there exists a third subfamily such that $\phi(x_\gamma) \to z$. This implies that $x_\gamma \to y - z$, and so A is compact. Suppose now that $F \subset \bar{D}$ is closed and that $x_\alpha + \phi(x_\alpha) \to z$, $x_\alpha \in F$. ϕ being compact, there exists $\{x_\beta\}$ such that $\phi(x_\beta) \to y$. Then $x_\beta \to z - y$; by continuity, $z - y + \phi(z - y) = z$. F is closed, so $z - y = \lim x_\beta$ belongs to F, which implies that $z \in (1 + \phi)(F)$. This means that $(1 + \phi)(F)$ is closed as desired.

Q.E.D.

3.48. Corollary:

$$\psi(\partial D) \text{ is closed}.$$

This property makes the statement "p is not in the closure of $\psi(\partial D)$" in most of the statements above, equivalent to "p is not in $\psi(\partial D)$", the same statement which appears in the finite dimensional case.

We leave to the reader the work (and the delight thereby engendered) of rewriting Propositions 3.41 and 3.43 for the case $\psi - 1 =$ compact, with the assumption $p \notin \psi(\partial D)$.

3.49. Corollary: Suppose D is a domain as in 3.43 and $\psi \in C(\bar{D})$ a map such that $\psi - 1$ is compact. If ψ maps D into a proper linear subspace of X, then $\psi(x) = \psi(-x)$ for some $x \in \partial D$.

Proof: Consider the map $\tilde{\psi}(x) = \psi(x) - \psi(-x)$. If $\tilde{\psi}(x) \neq 0$ when $x \in \partial D$, then by 3.44 $\tilde{\psi}(D)$ would cover a neighborhood of 0. But $\tilde{\psi}(D)$ is contained in every subspace containing $\psi(D)$; by hypothesis there is a proper one, without interior points.

Q.E.D.

3.50. Corollary: Let D be as above and $\psi \in C(\bar{D})$ such that $\psi - 1$ is compact. If $\psi(x)$ is never in the positive direction of $\psi(-x)$, $x \in \partial D$, then $\psi(x) = 0$ for some $x \in D$.

We shall define a notion of homotopy for compact mappings:

3.51. Definition: Two compact mappings $\phi_0, \phi_1 \in C(T)$ (T is any topological space) are said to be *compact homotopic* if there exists a compact mapping $F: I \times T \to X$, where $I = [0, 1]$, such that $F(0, x) = \phi_0(x)$, $F(1, x) = \phi_1(x)$.

3.52. Proposition: Let D be as above, $\phi_0 = \psi_0 - 1$, $\phi_1 = \psi_1 - 1$ two compact mappings $\phi_i \in C(\partial D)$. If ϕ_0 and ϕ_1 are compact homotopic under $F(t, x) = \phi_t(x) = \psi_t(x) - x$ and p is a point in X such that $p \neq \psi_t(x)$ for every t and every $x \in \partial D$, then

$$\deg(p, \psi_0, D) = \deg(p, \psi_1, D).$$

Proof: If F is compact, it may be approximated by finite dimensional mappings. The restrictions of such mappings also provide close approximations of ϕ_0 and ϕ_1, and then the proposition follows from the finite dimensional case.

Q.E.D.

N. Multiplicative Property and Generalized Jordan's Theorem for Banach Spaces

X is now a *Banach space*. Let $D \subset X$ be a bounded domain $\psi: \bar{D} \to X$, where $\psi - 1$ is compact.

3.53. Lemma:

$$\psi(\bar{D}) \text{ is bounded.}$$

Of course $\psi(\bar{D}) \subset \bar{D} + \phi(\bar{D})$, and $\phi(\bar{D})$ being compact, both $\bar{D}$ and $\phi(\bar{D})$ are bounded. Then $\psi(\bar{D})$ is bounded.

Q.E.D.

Since $\psi(\partial D)$ is closed (see 3.48), the set $\Delta = X - \psi(\partial D)$ is open and therefore has the form

$$\Delta = \bigcup_i \Delta_i$$

where the Δ_i are the components of Δ. Among these there is one and only one unbounded component, Δ_∞, because $\psi(\partial D)$ is bounded. Let $G = \bigcup_{i \neq \infty} \Delta_i$, and suppose furthermore that $g: \bar{G} \to X$ is a mapping such that $g - 1$ is compact.

3.54. Theorem: (multiplicative property) Under the hypotheses above, and if $p \notin g\psi(\partial D)$, then:

$$\text{(I)} \qquad \deg(p, g\psi, D) = \sum_{i \neq \infty} \deg(p, g, \Delta_i) \deg(\Delta_i, \psi, D).$$

Remark: We have used the notation $\deg(\Delta_i, \psi, D)$ as in section F (cf. 3.20): the justification for this comes from 3.41; 2.

Proof: First of all, as g is proper (3.47), it follows that $K = g^{-1}(p) \subseteq G$ is compact. Therefore from the covering $K \subset \bigcup_{i \neq \infty} \Delta_i$, we can select a finite family satisfying $K \subset \bigcup_{i=1}^{n} \Delta_i$. Thus in the expression (I) all but a finite number of terms vanish so that the sum is meaningful.

Moreover, if $g - 1$ is approximated very closely (uniformly over $\bar{G}$) by a finite dimensional mapping $\tilde{g} - 1$,

$$\deg(p, g, \Delta_i) = \deg(p, \tilde{g}, \Delta_i), \quad i \neq \infty,$$

as follows from the Definition 3.40.

Hence we can prove (I) as assuming $g - 1$ itself finite dimensional, and the general case will follow immediately.

Observe that if $\psi - 1$ is also finite dimensional we are done, because we then have just the finite dimensional result proved in 3.20. Thus the only thing to be proved is that ϕ may be approximated by finite dimensional mappings, for which (I) is already known.

When $\psi - 1$ is approximated closely, the composition $g\psi$ is also approximated, and the left member of (I) remains unchanged:

$$\deg(p, g\psi, D) = \deg(p, g\psi', D),$$

where ψ' is the mapping corresponding to a suitable approximation to $\phi = \psi - 1$.

Of course the terms $\deg(q, \psi, D)$ don't change either after the substitution of ψ' for ψ. The only difficulty arises when we consider the sets Δ_i, which obviously do change. But $K = g^{-1}(p)$ is *compact*, so the new sets Δ_i' will differ from the old ones by some (closed) sets, disjoint from K. By 3.41; 4 applied to these closed sets, the desired equality follows.

Q.E.D.

Suppose again that D is open and bounded and that $\psi: \bar{D} \to X$ with $\phi = \psi - 1$ compact.

3.55. Lemma: If $\psi: \bar{D} \to X$ is one-to-one, then $\psi^{-1} - 1: \psi(\bar{D}) \to X$ is compact.

Proof: It is easy to see that $\psi^{-1} - 1 = \phi \circ \psi^{-1}$ which yields the lemma. Q.E.D.

3.56. Lemma: ψ can be extended to $\Psi: X \to X$ in such a way that $\Psi - 1$ is still compact.

The proof will follow immediately from Proposition 3.58 below.

3.57. Theorem: (generalized Jordan's theorem) If D and D^* are bounded open sets in a Banach space X and there exists a homeomorphism $\psi: \bar{D} \to \bar{D}^*$ such that $(\psi - 1)(\bar{D})$ is compact, then the number of components of $X - D$ and $X - D^*$ is the same.

Proof: By Lemma 3.55, the inverse mapping ψ^{-1} is also of the form (identity) + (compact); thus the hypotheses are symmetric.

But by Lemma 3.56 it is also possible to assume that ψ and ψ^{-1} are restrictions of globally defined compact perturbations of the identity. The proof now is the same as that of 3.21, except for the fact that the appeals to 3.20 are replaced by references to 3.54. Q.E.D.

We now give a proof of Lemma 3.56. This lemma is an immediate consequence of the following generalization of the Tietze theorem due to J. Dugundji (An extension of Tietze's theorem, Pacif. Journal, Vol. 1, pp. 353–367 (1951)).

3.58. Proposition: Let A be a closed subset of a metric space X, and C a convex set in a locally convex T.L.S. E over the real or the complex field.

Then any continuous $f: A \to C$ has a continuous extension $F: X \to C$.

Proof: For each $x \in X - A$ choose an open B containing x such that diam $V_x \leqq \varrho(V_x, A)$. Then $\{V_x\}$ is an open covering of $X - A$ and since $X - A$ is paracompact there exists a locally finite refinement $\{U\}$, i.e., the U's are open and cover $X - A$, each $U \subset$ some V_x, and for each $x \in X - A$ there exists an open O_x containing x and disjoint from all but a finite number of the U's.

Let $U_0 \in \{U\}$ and define for $x \in X - A$

$$\lambda_{U_0}(x) = \varrho(x, X - U_0) / \sum_U \varrho(x, X - U).$$

Since $\varrho(x, X - U) > 0$ iff $x \in U$, and since each $x \in X - A$ is contained in some U we have $0 \leqq \lambda_{U_0}(x) \leqq 1$. For any $x \in X - A$, $\lambda_{U_0}|O_x$ has the form

$$\varrho(x, X - U_0) / \sum_{\substack{\text{finite} \\ \text{no. of } U\text{'s}}} \varrho(x, X - U),$$

and since each $\varrho(x, X - U)$ is continuous (because

$$|\varrho(x, X - U) - \varrho(y, X - U)| \leqq \varrho(x, y)),$$

$\lambda_{U_0}|O_x$ is continuous. Therefore λ_{U_0} is continuous on $X - A$ and $\lambda_{U_0}(x) = 0$ iff $x \notin U_0$.

Now for each U choose $a_U \in A$ such that $\varrho(a_U, U) < 2\varrho(A, U)$, and let the extension F be given by

$$F(x) = \sum_U \lambda_U(x) f(a_U) \quad \text{for} \quad x \in X - A,$$

and

$$F(x) = f(x), \qquad \text{for} \quad x \in A.$$

For each $x \in X - A$, $\lambda_U(x) = 0$ except for finitely many U's and since $\sum_U \lambda_U(x) = 1$, and $f(a_U) \in C$ it follows that $F(x) \in C$. If $x \in X - A$, $F|O_x$ is a finite sum of continuous functions and hence F is continuous on $X - A$. Since F is continuous on the interior of A by assumption, it only remains to show the continuity of F on the boundary of A.

Let $x_0 \in$ boundary A, and let $W \subset E$ be any convex open set containing the origin. Since f is continuous on A there exists an $\alpha > 0$ such that if $a \in A$ and $\varrho(x_0, a) < \alpha$ then $f(a) - f(x_0) \in W$. Let $O = \{x \in X : \varrho(x, x_0) < \alpha/6\}$. We will show that if $x \in O$, then $F(x) - F(x_0) \in W$.

Assume $x \in X - A$, $\varrho(x, x_0) < \alpha/6$ and $\varrho(x, a_U) < \alpha/2$. Then

$$\varrho(x_0, a_U) \leqq \varrho(x_0, x) + \varrho(x, a_U) < \frac{\alpha}{6} + \frac{\alpha}{2} < \alpha$$

implies $f(a_U) - f(x_0) \in W$. On the other hand assume $x \in X - A$, $\varrho(x, x_0) < \alpha/6$ and $\varrho(x, a_U) \geqq \alpha/2$. Then $\varrho(x, a_U) \geqq 3\varrho(x, x_0) \geqq 3\varrho(x, A)$. If $x \in U$ then

$$\varrho(x, a_U) \leqq \varrho(a_U, U) + \operatorname{diam} U < 2\varrho(A, U) + \operatorname{diam} U.$$

Since $U \subset V_x$ and $\operatorname{diam} V_x \leqq \varrho(V_x, C)$ we have

$$\operatorname{diam} U \leqq \operatorname{diam} V_x \leqq \varrho(V_x, A) \leqq \varrho(U, A).$$

Therefore $\varrho(x, a_U) \geqq 3\varrho(U, A) \leqq 3\varrho(A, x)$ contradicting the inequality above. Hence $\varrho(x, x_0) < \alpha/6$ and $\varrho(x, a_U) \geqq \alpha/2$ implies $x \notin U$ and therefore $\lambda_U(x) = 0$.

Finally for $x \in X - A$ and $\varrho(x, x_0) < \alpha/6$ we have

$$F(x) - F(x_0) = \sum_U \lambda_U(x) f(a_U) - f(x_0) = \sum_U \lambda_U(x)\,(f(a_U) - f(x_0))$$

and by the above, for each U either $\lambda_U(x) = 0$ or $f(a_U) - f(x_0) \in W$. Since the sum is actually a finite sum and $\sum_U \lambda_U(x) = 1$ with $0 \leqq \lambda_U(x) \leqq 1$ it follows that $F(x) - F(x_0) \in W$.

Q.E.D.

O. Fixed Point Theorems in Banach Spaces

Let X be a Banach space, $K \subset K$ a *convex* and *compact* set.

3.59. Lemma: Every continuous mapping $f: K \to K$ has a fixed point x such that: $f(x) = x$.

Proof: Since K is compact it is contained in some ball B around the origin. Each ball in a Banach space is a metric space, so by 3.58 we can assume that f is the restriction of a continuous mapping (also called f) from B into K. Clearly any fixed point of the extension is a fixed point of the original mapping.

Consider now the family of mappings

$$\psi_t = I - tf, \quad 0 \leqq t \leqq 1.$$

Since K is compact ψ_t depends continuously on t in the uniform topology (it suffices to observe that $t \to 0$ implies $tf \to 0$ uniformly).

The homotopy $F(t, x) \to tf(x)$ is *compact* because the set $\{ty; 0 \leqq t \leqq 1, y \in K\}$; is the continuous image of the compact set $[0, 1] \times K$ under the mapping $(t, y) \to ty$, and is therefore compact. That implies that

$$1 = \deg(0, \psi_0, B) = \deg(0, \psi_1, B)$$

which establishes at once the existence of a point $x \in B$ satisfying $\psi_1(x) = 0$, or $x - f(x) = 0$, in other words, a fixed point of f.

Q.E.D.

3.60. Proposition: (Schauder) Let A be a closed convex set contained in the Banach space X. Every compact mapping $f: A \to A$ has a fixed point.

Proof: Suppose $f(A) = \tilde{K}$. Let K be the closed convex hull of $\tilde{K}$. K is contained in A and the restriction $f|_K$ has a fixed point by the lemma above.

Q.E.D.

3.61. Proposition (Rothe): Let A be a convex bounded open set in some Banach space X. Suppose $\phi : \bar{A} \to X$ is compact and $\phi(\partial A) \subset A$. Then ϕ has a fixed point.

Proof: Let us suppose that $0 \in A$. Define p by

$$p(x) = \sup \{\lambda; 0 \leqq \lambda, \lambda x \in A\}.$$

Clearly p is continuous. Let $q(x) = \max \{p(x), 1\}$. Then q is also continuous. Hence the map $g(x) = x/q(x)$ is continuous, sends the whole space X into $\bar{A}$ and is the identity on $\bar{A}$. Moreover, the properties $q(x) = 1$ and $x \in \partial A$ are equivalent.

Let B be a ball containing $\bar{A}$ and $\phi(\bar{A})$.

The mapping $\tilde{\phi} = \phi \circ g : B \to B$ is compact (because ϕ is compact) and by Schauder's theorem has a fixed point $x = \tilde{\phi}(x)$. This point is easily seen to be in A, and thus it is a fixed point for ϕ.

Q.E.D.

3.62. Proposition: (Altman) Let A be any convex open bounded set in a Banach space X containing 0, and $\phi : \bar{A} \to X$ a continuous mapping satisfying

$$|x - \phi(x)|^2 \geqq |\phi(x)|^2 - |x|^2, \quad x \in \partial A.$$

Then ϕ has a fixed point.

Proof: For $0 \leqq t \leqq 1$ and, $x \in \bar{A}$ define the mapping

$$F(t, x) = t\phi(x).$$

F is easily seen to be compact.

For t fixed, write

$$d(t) = \deg (0, I - F(t, x), A).$$

Clearly $d(0) = 1$.

Furthermore, if it is supposed that $0 = F(t_0, x_0)$, for some $x \in \partial A$, $0 \leqq t_0 \leqq 1$, then

$$t_0\phi(x_0) = x_0,$$

or

$$|\phi(x_0)| = \frac{1}{t_0}|x_0|,$$

$$|\phi(x_0) - x_0|^2 = (1 - t_0)^2 |\phi(x_0)|^2 = \frac{(1 - t_0)^2}{t_0^2}|x_0|^2,$$

$$|\phi(x_0)|^2 - |x_0|^2 = |x_0|^2\left(\frac{1}{t_0^2} - 1\right) = |x_0|^2\frac{1 - t_0^2}{t_0^2}.$$

Using the hypothesis, we get

$$|x_0|^2\frac{(1 - t_0)^2}{t_0^2} \geqq |x_0|^2\frac{1 - t_0^2}{t_0^2}$$

or

$$(1 - t_0)^2 \geqq 1 - t_0^2,$$

which cannot occur for any $0 < t_0 \leqq 1$. Then the homotopy $F(t, x)$ is compact and avoids the point 0. Therefore

$$1 = d(0) = d(1)$$

which means that $\phi(x) = x$ has a solution for some $x \in A$.
Q.E.D.

As an application of homotopy invariance we shall obtain now a separation property.

3.63. Proposition: Let $K \subset X$ be bounded and closed; x_1, x_2 elements of X belonging to different components of $X - K$. If $F(t, x)$, $x \in K$ is a compact homotopy such that $F(0, x) = 0$, and if $\phi_t(x) = x - F(t, x)$ is such that $\phi_t(K)$ never contains x_1 or x_2 then x_1 and x_2 belong to different components of $X - \phi_1(K)$.

Proof: At least one of x_1 and x_2 belongs to a bounded component. Suppose it is x_1. Let Δ denote any component of $X - K$. Now compare

$$\deg(x_1, \phi_0, \Delta) \quad \text{and} \quad \deg(x_2, \phi_0, \Delta).$$

They are different. After the homotopy, they remain different, which implies that x_1, x_2 don't belong to the same component.
Q.E.D.

CHAPTER IV

Morse Theory on Hilbert Manifolds

Part 1

Part 2

This chapter consists of two parts: a discussion of the elementary properties of smooth manifolds modelled on general linear topological spaces and the theory of critical points of mappings defined on such manifolds (actually, when the L.T.S. is a Hilbert space). The two main references for the second part are (1) and (2) of the bibliography. Moreover, for the classical bibliography, references may be found in J. Milnor, *Morse Theory*, Ann. of Math. Studies, No. 51, Princ. (1963). We assume that the reader is familiar with finite dimensional differentiable manifolds (de Rham's book, Chern's lecture notes—Chicago—, Helgason's book). All the definitions and properties in Part 1 below are easy generalizations of the corresponding ones in the classical case.

Part 1

A. Manifolds

We recall that a mapping from an open set of a linear topological space with values in another L.T.S. is called *smooth* if it is differentiable infinitely many times in the Frechet sense.

4.1. Definition: If V is a L.T.S. and M a topological space, we shall define *a smooth V-structure on M* as a pair $(\mathscr{U}, \mathscr{I})$, where

1. $\mathscr{U} = \{U_\alpha\}$ is an open covering of M.
2. $\mathscr{I} = \{\phi_\alpha\}$ is a family of mappings, $\phi_\alpha : U_\alpha \to V$ such that if $U'_\alpha = \phi_\alpha(U_\alpha)$ then U'_α is open and
(a) $\phi_\alpha : U_\alpha \to U'_\alpha$ is a homeomorphism,
(b) $\phi_\alpha\phi_\beta^{-1} : \phi_\beta(U_\alpha \cap U_\beta) \to \phi_\alpha(U_\alpha \cap U_\beta)$ is smooth.

If M has been provided with a smooth V-structure, we shall say that M is a smooth manifold modelled on V. Every $\phi_\alpha \in \mathscr{I}$ will be called a *chart*, a *coordinate system* or a *map* on M, U_α being its *domain*.

Remark: If (b) is replaced by

(b′) $\phi_\alpha\phi_\beta^{-1}$ is k-times differentiable.

or by

(b″) $\phi_\alpha\phi_\beta^{-1}$ is analytic,

we obtain the concepts of *C^k-V-structure on M* and of *C^ω-V-structure*, respectively. In most cases, the V-structures are smooth ones (or *C^∞-V-structures*, as one says), and we shall restrict ourselves to this case. Parallel results for other cases (when true!) can be proved by the reader.

Examples:

1. Every open set $M \subset V$ can be modelled on V by means of the smooth (indeed analytic) V-structures provided by the family $\mathscr{J} = \{i\}$, where $i: M \to V$ is the inclusion.

2. Let Z be a discrete subgroup of V, and define $M = V/Z$, with the quotient topology. Consider all pairs V_α, V'_α, V'_α = open set of V, V_α = open set of M, such that the canonical projection $\pi: V \to M$ sends V'_α *homeomorphically* onto V_α. If $\mathscr{J}$ is defined as the family of inverses $\pi^{-1}: V_\alpha \to V'_\alpha$, then M together with $\mathscr{J}$ satisfies the requirements above; hence M has a C^∞ structure modelled on V, or, briefly, M is a C^∞ V-manifold. (If Z is not assumed to be discrete, $\mathscr{J}$ does *not* define a smooth structure nor even a C^1 structure. Why?)

3. We shall see later (4.51, (a)) that in every Hilbert space $\mathscr{H}$, the unit sphere $\{x \mid |x| = 1\}$ can be modelled on V, where V is any quotient $\mathscr{H}/\mathscr{H}_0$, $\mathscr{H}_0$ being an one dimensional subspace of $\mathscr{H}$.

4. New examples can be obtained from known ones by means of the two following procedures.

(a) if $G \subset M$ is open and M is a V-manifold (say $(\mathscr{U}, \mathscr{J})$), then G has an induced V-structure defined by the restrictions of the mappings in $\mathscr{J}$ to the sets $U \cap G$, $U_\alpha \in \mathscr{U}$.

(b) if M has a V-structure and L has a W-structure then $M \times L$ has a $V \times W$-structure in the obvious way.

B. Functions

Let M be a V-manifold and M' a V'-manifold. Let $\mathscr{I}$, $\mathscr{I}'$, ϕ, ϕ', U, U', define the manifold structures.

4.2. Definition: A mapping $f: M \to M'$ is called a *smooth mapping* if it is continuous and for every pair $\phi' \in \mathscr{I}'$, $\phi \in \mathscr{I}$, the mapping $\phi' f \phi^{-1}$—which is defined on some open set of V (precisely: $(\phi)^{-1}(U \cap f^{-1}(U'))$) and takes its values in V'—is smooth. The set of all smooth mappings $f: M \to M'$ will be denoted by *Hom* (M, M').

4.3. Definition: M and M' are *diffeomorphic* if there exist $f \in \text{Hom}(M, M')$ and $g \in \text{Hom}(M', M)$ such that $fg = id: M' \to M'$ and $gf = id: M \to M$.

If M'' is a V'' manifold, there exists a natural mapping

$$\text{Hom}(M, M') \times \text{Hom}(M', M'') \to \text{Hom}(M, M'')$$

defined by composition of mappings.

4.4. Definition: Let p be a point in M, U an open set of M containing p. A mapping $f \in \text{Hom}(U, M')$ is *horizontal at* p iff $(f\phi^{-1})'(\phi(p)) = 0$ for one (and hence for every) chart $\phi \in \mathscr{J}$ whose domain contains p. If f is a real-function horizontal at p, we shall also say that p is a *critical point* of f and that the real number $f(p)$ is a *critical level* of f.

Suppose now that ψ is another chart $\psi \in \mathscr{J}$ defined near p. By the chain rule applied to $(f\phi^{-1}) \circ (\phi\psi^{-1})$ we see that:

$$\begin{aligned}(f\psi^{-1})'(\psi(p)) &= ((f\phi^{-1})(\phi\psi^{-1}))'(\psi(p)) \\ &= [(f\phi^{-1})'(\phi\psi^{-1})(\psi(p))]\,[(\phi\psi^{-1})'(\psi(p))] \\ &= [(f\phi^{-1})'(\phi(p))]\,[(\phi\psi^{-1})'(\psi(p))].\end{aligned}$$

Thus $(f\phi^{-1})'(\phi(p)) = 0$ iff $(f\psi^{-1})'(\psi(p)) = 0$, because the mapping $(\phi\psi^{-1})' \times (\psi(p))$ is invertible (by the implicit function theorem), and this shows the definition of being horizontal at p to be independent of the chart chosen.

We shall henceforth write $C^\infty(M)$ for Hom (M, R), R with its standard structure. Note that $C^\infty(M)$ is a real algebra.

C. Tangent Vectors

Our aim is to define the tangent vector to a curve in a manifold at any point through which the curve passes. Let x be a point in the E-manifold M. Consider the set of all the smooth real functions defined on some open set containing x, and the equivalence relation $\sim$ on that set defined by

$$f \sim g \quad \text{iff} \quad f = g \quad \text{on some neighborhood of } x.$$

4.5. Definition: We shall define a *germ of smooth functions* at x to be any class of functions modulo the relation $\sim$. The set of germs at x will be denoted $G(x)$ and if f is a function, $\gamma(f)$ will denote its germ (i.e., the germ containing f).

It is clear that $G(x)$ is a real algebra. We want to consider tangent vectors at x. Roughly speaking, they will be "classes of curves going through x with the same velocity". We shall check both properties by means of the elements in $G(x)$.

4.6. Definition: A *curve* through $x \in M$ is a smooth mapping $p: J \to M$, where J is an open set of R such that $p(u) = x$ for some $u \in J$.

4.7. Definition: If p is a curve, $p(u) = x$, we shall define the tangent vector to p at $x = p(u)$ as the mapping $\left.\frac{dp}{dt}\right|_{t=u} : G(x) \to R$ defined by

$$\left.\frac{dp}{dt}\right|_{t=u} \gamma(f) = \left.\frac{d(f \circ p)}{dt}\right|_{t=u}.$$

The second member is meaningful because the mapping $f \circ p$ is a smooth real function of a real variable and for every such mapping g, we identify the linear mapping $g'(x): R \to R$ with the number $g'(x, 1)$.

4.8. Definition: The set TM_x of all the tangent vectors $\left.\frac{dp}{dt}\right|_{t=u}$ where p is a curve and $p(u) = x$, is called the *tangent space to M at x.*

A tangent vector $\left.\frac{dp}{dt}\right|_{t=u}$ is easily seen to be linear on $G(x)$, whence there is an injection $TM_x \subset (G(x))^*$, where $*$ denotes the algebraic dual space.

Of course we may assume, by changing the parameter, that every vector is of the form $\left.\frac{dp}{dt}\right|_{t=0}$, and we will do so hereafter. Now choose a chart ϕ at x. Every curve p with $p(0) = x$ induces (after cutting down its domain J, if necessary) a mapping $\phi \circ p: J \to E$. But then $(\phi \circ p)'(0): R \to E$ is linear. This means that $(\phi \circ p)'(0)$ may be identified with $e(p) = (\phi \circ p)'(0, 1)$. For every $\gamma(f) \in G(x)$ we have (cf. 1.7 for notation and the chain rule 1.14):

$$\begin{aligned}\left.\frac{dp}{dt}\right|_{t=0} \gamma(f) = (f \circ p)'(0, 1) &= (f \circ \phi^{-1} \circ \phi \circ p)'(0, 1)\\ &= (f \circ \phi^{-1})'(\phi(p(0)), (\phi \circ p)'(0, 1))\\ &= (f \circ \phi^{-1})'(\phi(x), e(p)).\end{aligned}$$

Thus

$$(1) \qquad \left.\frac{dp}{dt}\right|_{t=0} \gamma(f) = (f \circ \phi^{-1})'(\phi(x), e(p)).$$

Hence if p and q are two curves such that $e(p) = e(q)$, then the tangent vectors $\left.\frac{dp}{dt}\right|_{t=0}$, $\left.\frac{dq}{dt}\right|_{t=0}$ *coincide*. Therefore we can define a mapping $\phi_*(x)$: $TM_x \to E$ as follows.

4.9. Definition:

$$\phi_*(x)\left(\left.\frac{dp}{dt}\right|_{t=0}\right) = e(p).$$

The notation is chosen to emphasize the fact that the mapping $\phi_*(x)$ depends on the chart ϕ chosen (and of course on $x \in M$).

4.10. Definition: If $v \in TM_w$, the element $\phi_*(x)\, v$ of E is called the *coordinate* of v in the chart ϕ.

In the new notation, formula (1) becomes:

$$v\,(\gamma(f)) = (f \circ \phi^{-1})'\,(\phi(x), \phi_*(x)\, v) \tag{2}$$

4.11. Lemma: For every chart ϕ, the mapping $\phi_*(x) : TM_w \to E$ is one-to-one and onto. If ψ is another chart covering x, then

$$\phi_*(x) = S \circ \psi_*(x), \tag{3}$$

where S is the following automorphism of E: $S = (\psi \circ \phi^{-1})'\,(\phi(x))$.

Proof: From formula (2) we conclude that $\phi_*(x)\, v = \phi_*(x)\, w$ implies $v\gamma = w\gamma$ for every γ, or $v = w$. Hence $\phi_*(x)$ is one-to-one. Let e be an element of E, and define the curve $p(t) = \phi^{-1}(\phi(x) + te)$ (t small). It is very easy to prove that if $v = \left(\dfrac{dp}{dt}\right)_{t=0}$, then $\phi_*(x) v = e$. The formula (3) follows by standard computations.

A useful consequence of Lemma 4.11 is the fact that any $\phi_*(x)$ induces, by transport of structure from E, a structure of L.T.S. on TM_w, and that all $\phi_*(x)$ induce the same structure (this follows from formula (3) above). We sum up in the following proposition.

4.12. Proposition: Every TM_w has a canonical structure of L.T.S. given in such a way that for every chart ϕ around x, the mapping $\phi_*(x) : TM_w \to E$ characterized by

$$v\gamma\,(f) = (f \circ \phi^{-1})'\,(\phi(x), \phi_*(x)\, v) \tag{2}$$

is a L.T.S. isomorphism. Moreover, the linear structure so induced on TM_x has the following properties:

$$\left.\begin{aligned}(v + w)\,\gamma &= v\gamma + w\gamma \\ (\lambda v)\,\gamma &= \lambda \cdot v\gamma\end{aligned}\right\} \quad \text{for} \quad \gamma \in G(x),$$

or, in other words, coincides with that induced by the injection $TM_w \subset (G(x))$.*

The last part of the proposition is very easy, and is left to the reader.

The definitions make the following proposition obvious.

4.13. Proposition: Every $v \in TM_x$ is a *derivation* of the algebra $G(x)$ in R, i.e.

$$v(\gamma(f)\gamma(g)) = (v\gamma(f))\, g(x) + f(x) \cdot (v\gamma(g)).$$

Warning: The converse assertion (that every derivation is a tangent vector) is false—see below—unless E is finite dimensional.

D. Alternative Definitions of Tangent Vectors

We continue with the notation of the last section.

(a) *First alternative.*

Let $G(x)$ be the algebra of germs of smooth functions defined near $x \in M$. Choose a chart ϕ around x (sending x into $0 \in E$). Then every $\gamma \in G(x)$ is the germ of a function $\gamma = \gamma(f)$, and by the Taylor formula

$$f \circ \phi^{-1} = c + x' + s$$

where c is a real constant, x' is an element of E' and s is a second order—map (from a neighborhood of $0 \in E$ onto R). If ψ is another chart (also sending x into 0), and

$$f \circ \psi^{-1} = \hat{c} + \hat{x}' + \hat{s}$$

is the corresponding decompositon, then $c = \hat{c}$ and $\hat{x} = x' \circ T$, where $T = (\phi \circ \psi^{-1})'(0)$. This is clear after easy computation. T is an automorphism of E, both algebraic and topological.

Chosing ϕ, define $\bar{\phi}(f) = x'$. From the remark above, it follows that $\bar{\phi}(f) = \bar{\psi}(f) \circ [(\phi \circ \psi^{-1})'(0)]$. This means that the mappings $\bar{\phi} : G(x) \to E'$ induced by all charts ϕ are essentially the same. In particular it is true that a topology $\mathscr{C}$ on $G(x)$ may be defined as the weakest among those linear topologies for which $\bar{\phi}$ is continuous, when E' is supposed to be endowed with the weak topology of the duality (E, E'): this topology $\mathscr{C}$ is independent of ϕ.

Now define $D(x)$ as the set $D(x) = \{d\}$, where d are the following mappings:

(a) $d : G(x) \to R$, d is linear and continuous for $\mathscr{C}$;

(b) d is a derivation: $d(\gamma\mu) = d(\gamma) \cdot \mu(x) + \gamma(x)\, d(\mu)$.

Of course, if $f \circ \phi^{-1} = c + x' + s$, then $d(\gamma(f)) = d(\gamma(x' \circ \phi))$. For every $x' \in E'$ and $d \in D(x)$, write $d(x') = d(\gamma(x' \circ \phi))$. If $d(x') = d^*(x')$ for all

$x' \in E'$, then $d(\gamma) = d^*(\gamma)$ for all $\gamma \in G(x)$, or $d = d^*$. This means that if ϕ is chosen, there is an injection $D(x) \subset (E')^*$.

But the continuity required of the elements d implies that $D(x) \subset E \subset (E')^*$ One may now show immediately that $D(x) = E$.

$D(x)$ might be called the tangent space at x, and all the theory based upon this choice.

(b) *Second alternative.* Again we use the same notation.

Using equality (1) of section C, we may define the tangent vectors by means of their coordinates in all charts. In fact, if e is the coordinate of γ in the chart ϕ, then its coordinate in ψ is

[*] $$g = (\psi \circ q^{-1})' (\phi(x), e).$$

That means that γ can be identified with the class of all pairs (ϕ, e), (ψ, g), ... where ϕ (resp. ψ, ...) is a chart and $e \in E$ ($g \in E$, ...), satisfying [*]. Call V_x the set of all such classes. The correspondence: class of $(\phi, e) \rightarrow e$ defines, for a given ϕ, a map that identifies V_w with E, and we have a structure of L.T.S. for V_w.

Of course if we define a mapping

$$\text{class of } (\phi, e) \rightarrow \left.\frac{dp}{dt}\right|_{t=0}$$

where p is the curve defined as $p(t) = \phi^{-1}(\phi(x) + te)$, ($t$ small), then we obtain a one-to-one-onto correspondence between the elements in V_x and the tangent vectors to smooth curves. On the other hand, if we define the mapping

$$\text{class of } (\phi, e) \rightarrow d$$

where $d \in D(x)$ is defined by $d(\gamma(f)) = (f \circ \phi^{-1})' (\phi(x), e)$, then we obtain a correspondence between elements in V_x and elements in $D(x)$.

This proves that both approaches lead essentially to the same result.

As a final remark, let us observe that:

—In the first definition of tangent space as the set of tangent vectors to curves, addition is easily defined (by means of $TM_x \subset (G(x))^*$), but not the topology of TM_x. On the other hand, the geometrical meaning is very illuminating.

—In the second definition (alternative *a*), everything is natural (algebraic topological structure), except the meaning.

—The third definition (*b*) is neither natural nor meaningful.

E. More on Linear Topology

This section is used only in section K, and not in its full generality, but only in a very particular case.

Let E be any linear topological space. We shall denote by End (E) the space of continuous linear mappings $u: E \to E$.

4.14. Lemma: A linear (and Hausdorff if E is Hausdorff) topology for End (E) is defined by uniform convergence on bounded sets: a fundamental system of neighborhoods of the origin for this topology is the family of sets $L(K, V) = \{u \in \text{End}(E);\ u(K) \subset V\}$, where K ranges over the family of bounded sets of E and V over the neighborhoods of $0 \in E$.

Suppose now that E and $\hat{E}$ are two linear topological spaces, and consider the space $B(E, \hat{E})$ of continuous bilinear forms on $E \times \hat{E}$: $B(E, \hat{E}) = \{P; P: E \times \hat{E} \to R, P \text{ is bilinear and continuous}\}$. (The continuity of P is in the product topology.)

4.15. Lemma: $B(E, \hat{E})$ can be made into a Hausdorff L.T.S. by the topology of uniform convergence on bounded sets. A fundamental system of neighborhoods of $0 \in B(E, \hat{E})$ is provided by the sets $B_a(K, \hat{K}) = \{P \in B(E, \hat{E});\ P(K \times \hat{K}) \subset [-a, +a]\}$, where K and $\hat{K}$ range over the family of bounded sets of E and $\hat{E}$, respectively, and a over the positive numbers.

4.16. Lemma: The *bounded sets* of $B(E, \hat{E})$ are those sets D for which $D(K \times \hat{K})$ (defined as $\{P(x, y);\ P \in D,\ x \in K,\ y \in \hat{K}\}$) is bounded for every pair, K, $\hat{K}$ of bounded sets of E and $\hat{E}$, respectively.

The proof is left to the reader.

A subset $S \subset B(E, \hat{E})$ is called *equicontinuous* (or, better, *equicontinuous at the origin*) if for every neighborhood N of $0 \in R$, there exists a neighborhood N' of $(0, 0) \in E \times \hat{E}$ such that all P in S satisfy $P(N') \subset N$.

4.17. Lemma: *All equicontinuous sets in $B(E, \hat{E})$ are bounded.*

The proof follows immediately from (4.16). Suppose that $N = [-1, +1]$ and that $K \times \hat{K} \subset \lambda N'$ (definition of bounded sets); then $|P(x, y)| \leqq \lambda$ for all $P \in S$, $x \in K$, $y \in \hat{K}$.

There is a natural way to make End $(E) \oplus$ End $(\hat{E})$ operate on $B(E, \hat{E})$; the following lemma establishes this.

4.18. Lemma: The mapping j: End $(E) \times$ End $(\hat{E}) \to$ End $(B(E, \hat{E}))$ defined by $([j(u \oplus v)]P)(x, y) = P(ux, vy)$ is *bilinear*.

We note the following relation involving the mapping j:

4.19. Proposition: *Suppose that both E and $\hat{E}$ are locally convex. The mapping j is continuous iff the converse of 4.17 holds.*

Continuity of j from the Converse of 4.17

Let S be the neighborhood of $0 \in \text{End}(B(E, \hat{E}))$ defined by $S = \{u; uK \subset V\}$, where K is a bounded set of $B(E, \hat{E})$ and V a neighborhood of $0 \in B(E, \hat{E})$. Assume that $V = \{P; P(L \times \hat{L}) \subset [-a, +a]\}$, where L and $\hat{L}$ are bounded and a is a positive real number. Since we assume the converse of 4.17, K is also equicontinuous. Let N, $\hat{N}$ be neighborhoods of 0 in E and $\hat{E}$ respectively, such that if $P \in K$, then $P(N \times \hat{N}) \subset [-a, +a]$.

Define the neighborhoods

$$W = \{u \in \text{End}(E); u(L) \subset N\}$$

$$\hat{W} = \{u \in \text{End}(\hat{E}); u(\hat{L}) \subset N\}.$$

Then, if $u \in W$, $v \in \hat{W}$, it follows for every $P \in K$ that:

$$(j(u, v)P) L \times \hat{L} = P(uL \times v\hat{L}) \subset P(N \times \hat{N}) \subset [-a, +a],$$

which proves that $j(W \times \hat{W})$ is contained in S. Hence j is continuous.

Converse of 4.17 From the Continuity of j

Let K be a bounded set in $B(E, \hat{E})$; $a \in R$, $a > 0$. Choose two bounded sets L, $\hat{L}$ in E, $\hat{E}$ respectively, both different from $\{0\}$, and define

$$V = \{P \in B(E, \hat{E}); P(L \times \hat{L}) \subset [-a, +a]\}.$$

Clearly V is a neighborhood of $0 \in B(E, \hat{E})$. Now consider the set

$$S = \{u \in \text{End}(B(E, \hat{E})); uK \subset V\}.$$

S is a neighborhood of $0 \in \text{End}(B(E, \hat{E}))$. Since we assume that j is continuous, there exist N, $\hat{N}$ in End (E) and End $(\hat{E})$ such that $j(N \times \hat{N}) \subset S$, and we may suppose that N and $\hat{N}$ are of the form

$$N = \{u \in \text{End}(E); uL_1 \subset T_1\}$$

$$N = \{u \in \text{End}(\hat{E}); uL_2 \subset T_2\},$$

where L_i $(i = 1, 2)$ are bounded and T_i neighborhoods of 0. We may also suppose that $L_1 \supset L$ and $L_2 \supset \hat{L}$. But the inclusion $j(N \times \hat{N}) \subset S$ means that

$$j(n_1, n_2) P \in V \quad \text{for all} \quad n_i \in N_i, \quad P \in K \quad (i = 1, 2),$$

or

$$(j(n_1, n_2) P) L \times \hat{L} \subset [-a, +a]$$

and, finally

$$P(n_1 L \times n_2 \hat{L}) \subset [-a, +a] \quad \text{for all} \quad n_i \in N_i, \quad P \in K \quad (i = 1, 2).$$

Since $L \neq \{0\}$ and $\hat{L} \neq \{0\}$ this implies, using the Hahn-Banach Theorem, that

$$P(T_1 \times T_2) \subset [-a, +a].$$

Hence K is equicontinuous.

We shall now discuss some special cases for which the converse of 4.17 holds.

4.20. Definition: A linear topological space is called a *Baire space* if whenever the union of a countable family of closed subsets covers the whole space, then at least one of the subsets has non-empty interior.

The classical Baire Category Theorem (see Kelley or Boubaki) implies the following proposition.

4.21. Proposition: *Every Frechet space is a Baire space.*

Moreover, the statement below follows easily from the definition.

4.22. Proposition: *Every locally convex linear topological space which is a Baire space is a barrelled space.*

(Cf. Boubaki, E.V.T., Chapter III, § 1, Prop. 1.)

4.23. Proposition: *If E and $\hat{E}$ are locally convex linear topological spaces and $E \times \hat{E}$ is a Baire space, then a subset of $B(E, \hat{E})$ is equicontinuous iff it is bounded (and consequently, the map j in 4.18 is continuous).*

Preliminary remark: If $E \times \hat{E}$ is a Baire space, then both E and $\hat{E}$ are Baire spaces (hence barrelled spaces).

Proof:

Suppose that $D \subset B(E, \hat{E})$ is bounded and that $N = [-a, +a]$, $a \in R$, $a > 0$. The set $G = \{(x, y) \in E \times \hat{E};\ P(x, y) \in N \text{ for all } P \in D\}$ is clearly *closed.* Moreover, D being bounded, for every pair (x, y) there exists a positive integer λ such that $P(x, y) \in \lambda N$ for all $p \in D$. Therefore $E \times \hat{E} = \bigcup nG$, n = positive integer. Since $E \times \hat{E}$ is a Baire space, some nG must have non-empty interior; hence $G = \dfrac{1}{n}(nG)$ has non-empty interior: call this interior U and choose $(\alpha, \beta) \in U$.

Now consider the set

$$G_1 = \{x \in E;\ P(x, \beta) \in N, \quad \text{for all} \quad P \in D\}.$$

$G_1 \subset E$ is clearly a *barrel*, and therefore (by our preliminary remark) a neighborhood of $0 \in E$. The same argument applies to $\hat{E}$ and so we conclude that there exist neighborhoods G_1, $\hat{G}_1$ of $0 \in E$ and $0 \in \hat{E}$ respectively such that

$$\text{(1)} \qquad \begin{aligned} &P(\alpha, y) \in N, \quad \text{for all} \quad P \in D \quad \text{and all} \quad y \in \hat{G}_1 \\ &P(x, \beta) \in N, \quad \text{for all} \quad P \in D \quad \text{and all} \quad x \in G_1. \end{aligned}$$

Furthermore, we may find new neighborhoods $V \subset G_1$, $\hat{V} \subset G_1$ such that $(\alpha + V) \times (\beta + \hat{V}) \subset G$; in particular

$$\text{(2)} \qquad P((\alpha + V) \times (\beta + \hat{V})) \subset N \quad \text{for all} \quad P \in D.$$

Suppose finally that $(x, y) \in V \times \hat{V}$. Then, for every $P \in D$

$$\begin{aligned} P(x, y) &= P(x + \alpha, y + \beta) - P(a, \beta) - P(x, \beta) - P(\alpha, y) \\ &\in P((\alpha + V) \times (\beta + \hat{V})) + P(\alpha, \hat{V}) + P(V, \beta) - P(\alpha, \beta). \end{aligned}$$

From (1) and (2) we now get:

$$\text{(3)} \qquad P(x, y) \in 3N - P(\alpha, \beta), \quad \text{for all} \quad P \in D.$$

Let us observe now that since D is bounded, the set $\{P(\alpha, \beta);\ P \in D\}$ is also bounded and hence contained in some qN, q a positive integer.

Therefore, by (3):

$$\{P(x, y);\ P \in D, x \in V, y \in \hat{V}\} \subset (q + 3)\, N,$$

or

$$\text{(4)} \qquad \{P(x, y);\ P \in D, x \in W, y \in \hat{W}\} \subset N,$$

where W and $\hat{W}$ are neighborhoods satisfying $(q + 3)\, W \subset V$, $\hat{W} = \hat{V}$. The formula (4) shows that D is equicontinuous, as desired. As a corollary, we obtain the following statement.

4.24. Corollary: *If E and $\hat{E}$ are locally convex spaces and $E \times \hat{E}$ is a Baire space, then the mapping j: $End(E) \oplus End(\hat{E}) \to End(B(E, \hat{E}))$ is continuous.*

F. More on Elementary Calculus

Let M be a topological space and F a L.T.S. Denote by $C(M, F)$ the set of continuous mappings from M into F. Clearly $C(M, F)$ has a natural linear structure. Suppose now that Z is another L.T.S. Then there is a natural identification

$$C(M, F) \oplus C(M, Z) = C(M, F \oplus Z)$$

which is a linear isomorphism (defined by $(\phi \oplus \psi)(x) = \phi(x) \oplus \psi(x)$). Suppose now that $u: F \to Z$ is a continuous mapping. u induces a map

$$u' : C(M, F) \to C(M, Z)$$

defined by composition $u'(\phi) = u \circ \phi$.

Our aim is to consider the case in which M is a smooth manifold and to deal with smooth mappings.

Assume that M is a smooth manifold.

4.25. Lemma: *If* $\phi \in C(M, F)$ and $\psi \in C(M, Z)$ are *smooth, then so is* $\phi \oplus \psi$ *and furthermore*

$$\delta(\phi \oplus \psi)(x, h) = \delta\phi(x, h) \oplus \delta\phi(x, h).$$

The proof follows from the remark that if $u \in C(M, F)$ and $v \in C(M, Z)$ are both $o(x)$, then so is $u \oplus v$.

Let now Z, F and G be three L.T.S.

4.26. Lemma: *Every continuous bilinear mapping* $u: Z \times F \to G$ *is differentiable, and moreover its derivatives are*

$$\delta u\,[(x, y); (h, k)] = u(x, k) + u(h, y)$$

$$\delta^2 u\,[(x, y), (x', y'); (h, k)] = u(x', k) + u(h, y')$$

$$\delta^n u \equiv 0 \quad \text{if} \quad n \geqq 3.$$

The proof follows from elementary remarks and the formula

$$u(x + h, y + k) - u(x, y) = \{u(x, k) + u(h, y)\} + u(h, k)$$

(see also "Quadratic Forms," in Chapter I).

Assume again that M is a smooth manifold, and that F, Z are *locally convex* L.T.S. such that in $B(F, Z)$ every bounded set is equicontinuous.

4.27. Proposition: *If* $\phi \in C(M, \text{End}(F))$ *and* $\psi \in C(M, \text{End}(Z))$ *are differentiable and we call* j *the natural mapping* $\text{End}(F) \oplus \text{End}(Z) \to \text{End}(B(F, Z))$, *then the mapping* $\beta \in C(M, \text{End}(B(F, Z))$ *defined by* $\beta = j \circ (\phi \oplus \psi)$ *is also differentiable.*

Proof: By Lemma 4.25, $\phi \oplus \psi \in C(M, \text{End}(F) \oplus \text{End}(Z))$ is differentiable. But by Proposition 4.23, j is continuous, hence (Lemma 4.26, above) differentiable. Therefore β, as a composition of two differentiable mappings is itself differentiable.

4.28. Corollary: *Proposition 4.27 holds under the hypothesis that* F *and* Z *are locally convex and* $F \times Z$ *is a Baire space.*

4.29. Notation: *If* $\phi \in C(M, \text{End}(F))$ *and* $\psi \in C(M, \text{End}(Z))$, *the mapping* $\beta = j \circ (\phi \oplus \psi) \in C(M, \text{End}(B(F, Z)))$ *defined in 4.27 will be called* $s(\phi, \psi)$. *With this notation, Proposition 4.27 reads: if* ϕ *and* ψ *are smooth, so is* $s(\phi, \psi)$.

G. A Short Outline of Smooth Linear Bundles

The reference is: Lang, *Introduction to differentiable manifolds.*

(a) *Definition*

Let M be a smooth E-manifold.

4.30. Definition: A smooth linear bundle on M consists of the following objects:

1. a space X;
2. a surjective mapping $\pi: X \to M$ (called the *projection*);
3. an L.T.S. structure on every set $\pi^{-1}(x)$, $x \in M$, (the set $\pi^{-1}(x)$ is called the *fiber* of the bundle over X, and denoted by X_x);
4. a linear topological space F;
5. an open covering $\{U, V, \ldots\}$ of M, and
6. for every U in this covering, a map

$$\tau_U : \pi^{-1}(U) \to U \times F;$$

These maps must satisfy:

7. the maps $\tau_U, \tau_V, \ldots,$ commute with the projections, i.e., the diagram:

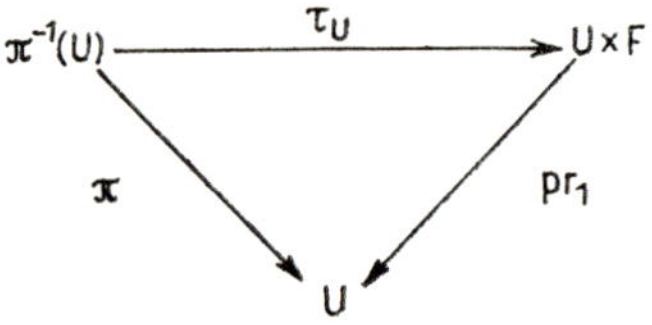

is commutative; (where $pr_1(x, e) = x$)

8. the restrictions $(\tau_U)_x : X_x \to \{x\} \times F$ are linear isomorphisms (here we assume that $\{x\} \times F$ and F are identified);
9. if U and V are two members of the covering, the map

$$\tau_{UV} : U \cap V \to \text{End}\,(F)$$

defined by

$$\tau_{UV}(x) = (\tau_U)_x\,[(\tau_V)_x]^{-1}, \quad x \in U \cap V$$

is a *smooth mapping* (End (F) has the topology defined in 4.14).

Examples

1. Let F be any L.T.S. Take $X = M \times F$, $\pi(m,f) = m$, the covering consisting only of the set M, and τ_M to be the identity. This is a smooth linear bundle, called the *trivial bundle.*

2. The tangent bundle of a manifold M, whose description appears in section **H** below.

Remark: Sometimes, for short, the expression "let X be a bundle" is used. The space F is called the *typical fiber* of the bundle (in the case of the tangent bundle, it coincides with the space on which M is modelled).

The notions of *sub-bundle of a bundle, direct sum of two bundles* and *homomorphism from one bundle into another* may be defined in the standard way.

Assume that the covering $\{U\}$ consists of domains of charts $\{\phi\}$. Then the mappings:

$$\pi^{-1}(U) \xrightarrow{\tau_U} U \times F \xrightarrow{\phi \times id} \phi(U) \times F \subset E \times F$$

are *charts* for an $E \times F$ manifold structure on X.

Remark: Bundles of class C^k and analytic bundles are similarly defined.

4.31. Definition: A *section* of a bundle X is any mapping $s : M \to X$ such that $\pi \circ s = id$.

(b) *A construction*

Remark: Here again, the full generality will not be necessary. The reader may assume that all spaces are Hilbert spaces and hence obtain the propositions more easily.

We shall give a procedure for constructing new bundles from known ones. Our procedure is suggested by the general description given in Lang (III, § 4).

Assume that M is a smooth E-manifold (E is not necessarily locally convex).

Let X and $\hat{X}$ be bundles on M; F and $\hat{F}$ their typical fibers, π and $\hat{\pi}$ their projections and τ_U, $\hat{\tau}_V$ their structural mappings. Define the set $B(X, \hat{X})$ to be

(a) $$B(X, \hat{X}) = \bigcup B(X_x, \hat{X}_x), \quad x \in M,$$

where $B(X_x, \hat{X}_x)$ denotes, as above, the space of bilinear forms on $X_x \times \hat{X}_x$.

We may assume that the coverings of M defining X and $\hat{X}$ are the same, replacing the original ones by the intersections $U \cap \hat{U}$, if necessary. Let

(b) $$\mathscr{A} = \{W, Q, \ldots\}$$

denote this covering. Now consider:

(c) $$p : B(X, \hat{X}) \to M,$$

the mapping defined as: if $u \in B(X_x, \hat{X}_x)$, then $p(u) = x$;

(d) for every $W \in \mathscr{A}$, define

$$\lambda_W : p^{-1}(W) \to W \times B(F, \hat{F})$$

as follows: for every $x \in W$, the maps $(\tau_W)_x$ and $(\hat{\tau}_W)_x$ give an identification of $X_x \times \hat{X}_x$ with $F \times \hat{F}$. Then every $u \in B(X_x, \hat{X}_x)$ induces an element $\tilde{u} \in B(F, \hat{F})$, which defines λ_W. In other words, if $u \in B(X_x, \hat{X}_x)$, then

$$\lambda_W(u) = (x, \tilde{u}) = (x, u \cdot [(\tau_W)_x^{-1}, (\hat{\tau}_W)_x^{-1}]).$$

The reader will verify that these objects satisfy properties (7) and (8) of the definition of bundle.

Let $\lambda_{QW} : Q \cap W \to \operatorname{End}(B(F, \hat{F}))$ denote the mappings $\lambda_{QW}(x) = (\lambda_Q)_x \times [(\lambda_W)_x]^{-1}$, where $Q, W \in \mathscr{A}$.

We see that this construction leads to a smooth fiber bundle as long as these maps λ_{QW} are smooth (condition (9)). But it is easily verified that

$$\lambda_{QW} = s(\tau_{QW}, \hat{\tau}_{QW}),$$

with the notation of 4.29.

4.32. Proposition: *Let X, $\hat{X}$ be two smooth fiber bundles on M (any smooth manifold), F and $\hat{F}$ their typical fibers. If both F and $\hat{F}$ are locally convex and in $B(F, \hat{F})$ every bounded set is equicontinuous, then the objects $B(X, \hat{X})$, p, $\mathscr{A}$ and λ_W described in* (a), (b), (c) *and* (d) *define a smooth linear bundle on M (its typical fiber being $B(F, \hat{F})$).*

Proof: Obvious from the fact that the hypothesis on $B(F, F)$ implies (by 4.28, 4.29) that the mappings $\lambda_{QW} = s(\tau_{QW}, \hat{\tau}_{QW})$ are smooth.

Remarks (1): Note that no assumptions are made on the local topology of M, but on the fibers of the bundles considered.

(2): Observe that the differentiability (or continuity by 4.26) of j: End (F) $\oplus$ End $(\hat{F}) \to$ End $(B(F, \hat{F}))$ is the weakest condition that can be imposed *on the fibers* in order to obtain the statement: if ϕ and ψ with values in End (F) and End $(\hat{F})$ are differentiable, then $j \circ (\phi \oplus \psi)$ is also differentiable. But the continuity of j is equivalent (see 4.19) to the property: every bounded set in $B(F, \hat{F})$ is equicontinuous. Hence we conclude that this latter property of $B(D, \hat{F})$ is a natural condition to impose in order to obtain the validity of the last proposition.

Nevertheless, an exception occurs in the case $X = \hat{X} =$ trivial bundle, where the proposition holds for any fiber.

(3): Suppose that $\hat{X}$ is the trivial bundle with fiber $\mathscr{R}$, the real numbers. Then $B(F, \hat{F}) = B(F, \mathscr{R}) = F'$, the topological dual of F provided with the strong topology.

Here the condition on F' about the bounded sets is equivalent (see Bourbaki, EVT Chap. III, § 3, Ex. 6) to: *the completion of F is barrelled.* Therefore:

4.33. Corollary: *Every linear bundle such that the completion of its fiber is a barrelled space has s smooth dual.*

Once again the requirement on the fiber is "necessary" (see Bourbaki, loc. cit.).

(4): Perhaps the reader cannot resist the temptation to define a tensor product of smooth linear bundles. The following might be a way: $X \otimes \hat{X}$ is defined by fibers as $(X \otimes \hat{X})_x =$ subspace generated in the dual $(B(X_x, \hat{X}_x))'$, of $B(X_x, \hat{X}_x)$ by the image of $X_x \times \hat{X}_x$ under the canonical mapping $X_x \times \hat{X}_x \to (B(X_x, \hat{X}_x))'$. This object is a smooth bundle provided that in $B(F, \hat{F})$ and in $(B(F, \hat{F}))'$ (the latter with the strong topology) every bounded set is equicontinuous.

This is always the case if F and $\hat{F}$ are spaces of type $(\mathscr{DF})$, for example.

Of course when the product $X \otimes \hat{X}$ exists, it has the standard universal property.

H. The Tangent Bundle

Throughout this section, M will denote a smooth E-manifold. We shall define a smooth linear bundle on M, the fiber being E, called the *tangent bundle* of M (notation: $T(M)$).

4.34. Definition: Let $T(M) = \bigcup_{x \in M} TM_x$, where TM_w is the tangent space to M at x; let $\pi : T(M) \to M$ be the projection $\pi(\gamma) = x$ if $\gamma \in TM_x$. Let $\mathscr{A} = \{U, V, \ldots\}$ be the open covering of M by the domains of charts $\phi, \psi, \ldots$ Put $T(\mathcal{O}) = \bigcup_{x \in \mathcal{O}} TM_x = \pi^{-1}(\mathcal{O})$ for every $\mathcal{O} \subset M$ and define $u_U : T(U) \to U \times E$ as follows: if $\gamma \in TM_x$, $u_U(\gamma) = (x, \phi_*(x)\gamma)$ (notation of 4.9). The system $T(M)$, π, $\{U, V, \ldots\}$, $\{u_U, u_V, \ldots\}$ defines a smooth linear bundle with fiber E. We call it the *tangent bundle of M*.

Of course the diagram

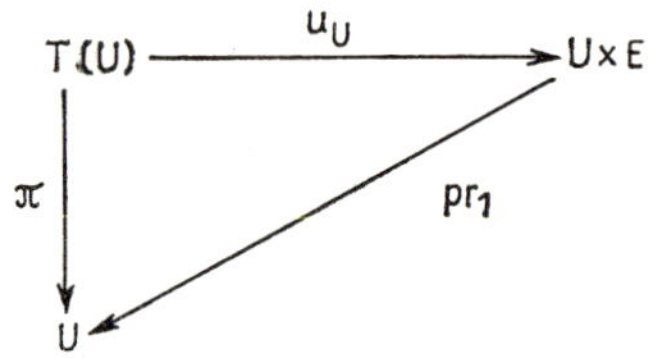

(where $pr_1(x, e) = x$) is commutative.

Consider now two charts ϕ, ψ, with domains U and V, respectively. The composition of the mappings:

$$(U \cap V) \times E \xrightarrow{u_U^{-1}} T(U \cap V) \xrightarrow{u_V} (U \cap V) \times E$$

is:

$$u_U(u_V)^{-1}(x, e) = (x, \phi_*(x)[\psi_*(x)]^{-1} e),$$

as follows immediately from the definition of u_U and u_V. But then, since the mapping

[*] $$\tau_{U,V} : x \to \phi_*(x)[\psi_*(x)]^{-1}$$

(from $U \cap V$ into End (E)) is a smooth mapping the mappings u_U make $T(M)$ into a smooth linear bundle and consequently into a smooth $E \times E$ manifold. (If M is a manifold of class C^k, $T(M)$ is a manifold of class C^{k-1}; if M is analytic, so is $T(M)$.)

Recall that in 4.31 the notion of *section* of bundle was defined.

4.35. Definition: A *vector field* on M is any section of $T(M)$.

A vector field may be of class C^k, $k \geqq 0$, C^∞ (analytic if M is analytic).

4.36. Definition: The set of vector fields of class C^k, $k < \infty$ will be denoted by Ω_k (or $\Omega_k(M)$, if necessary), and that of those of class C^∞ by (or $\Omega(M)$).

Vector fields defined only on an open set of M will often be considered.

Lifting of mappings. Suppose now that M and N are smooth manifolds (modelled on E and F, respectively) and that $f: M \to N$ is also smooth. Then a mapping f_* completing the diagram

$$\begin{array}{ccc} T(M) & \xrightarrow{f_*} & T(N) \\ \downarrow{\scriptstyle \pi} & & \downarrow{\scriptstyle \pi} \\ M & \xrightarrow{f} & N \end{array}$$

may be defined by means of

[**] $$f_*\left(\left.\frac{dp}{dt}\right|_{t=0}\right) = \left.\frac{d(f \circ p)}{dt}\right|_{t=0}.$$

Sometimes we shall write $f_*(x)$ for the restriction of f_* to TM_x:

[***] $$f_*(x): TM_x \to TN_{f(x)}.$$

4.37. Definition: The mapping $f_*(x)$ defined in [***] is called the *differential* of f at x.

4.38. Proposition: If $f: M \to N$ is C^1, then $f_*(x): TM_w \to TN_x$ is a continuous linear mapping for every $x \in M$. f is one-to-one on some neighborhood of x iff $f_*(x)$ is one-to-one. $f(U)$ covers a neighborhood of $f(x)$ for every neighborhood U of x iff $f_*(x)$ is onto.

The proof, which uses the implicit function theorem, is left to the reader.

Observe that the differential at x of a chart $\phi: U \to E$ (where $U \subset M$) is the mapping denoted by $\phi_*(x)$ in definition 4.9 (assume the identification $TE_y = E$ induced by the identity).

The mapping f_* can be computed as the derivative of a mapping when charts are chosen around x and $f(x)$. In fact, suppose that ϕ and ψ are such charts. Then, if $v = \left.\frac{dp}{dt}\right|_{t=0} \in TM_x$, and $y = f(x)$, we have

$$\begin{aligned} \psi_*(y)\,(f_*(x)\,v) &= \psi_*(y)\left(\left.\frac{d(fp)}{dt}\right|_{t=0}\right) \\ &= [(\psi f p)'(0)]\,(1) \\ &= [(\psi f \phi^{-1}\phi p)'(0)]\,(1) \\ &= [(\psi f \phi^{-1})'(\phi p(0))]\,((\phi p)'(1)) \\ &= [(\psi f \phi^{-1})'(\phi(x))]\,(\phi_*(x)\,v). \end{aligned}$$

Hence the following diagram in which $u = [(\psi f \phi^{-1})' (\phi(x))]$ commutes.

$$\begin{array}{ccc} E & \xrightarrow{u} & F \\ \uparrow{\scriptstyle \phi_*(x)} & & \uparrow{\scriptstyle \psi_*(f(x))} \\ TM_x & \xrightarrow{f_*(x)} & TN_{f(x)} \end{array}$$

I. Ordinary Differential Equations

In this section, H will denote a smooth manifold modelled on a *Banach space E*. Similar results may be obtained for manifolds modelled on Montel spaces (see E. Dubinsky, "Differential equations and differential calculus in Montel spaces", Trans. Am. Math. Soc., Vol. 110, No. 1 (1964)). The propositions below seem to have appeared for the first time in Michal and Elconin, "Completely integrable differential equations in abstract spaces", Acta Math., Vol. 68 (1937) pp. 71–107. Most proofs are omitted here: reference to the book of Serge Lang will provide them.

4.39. Notation: If p is a curve in M, we shall write $p'(u)$ instead of $\left.\dfrac{dp}{dt}\right|_{t=u}$.

4.40. Proposition: Let $p_0 \in M$, v be a vector field defined on some neighborhood of p_0. Consider the equation

$$\text{(I)} \qquad \begin{aligned} p'(t) &= v\,(p(t)) \\ p(0) &= p_0 \end{aligned}$$

where the unknown p is a curve in M, its parameter running over some interval containing $0 \in R$. If v is of class C^1, the equation (I) has solutions and two such solutions agree in the intersection of their intervals of definition.

The statement being local, M may be replaced by E itself. The proof is then a slight modification of the Picard's proof from the corresponding statement in the finite dimensional case. (Observe that the special form of the equation (I) implies immediately that if a solution is C^p, then it is necessarily C^{p+1}). As in the finite dimensional case, the above proposition implies the existence of maximal integral curves, in the following sense:

4.41. Proposition: *Suppose v is a C^n vector field defined on the manifold M; then, for every $p \in M$, there exists a smooth curve $\sigma\,(t, p)$ such that*

(a) *$\sigma(t, p)$ is defined for t belonging to some interval $(t_-(p), t_+(p))$, containing $0 \in R$, and is of class C^{n+1} there.*

(b) *$\sigma(0, p) = p$ for every p.*

(c) *$\sigma(t, p)$ satisfies the equation*

$$\left.\frac{d\sigma(t, p)}{dt}\right|_{t=u} = v(\sigma(u, p)).$$

(d) *Given $p \in M$, there is no C^1 curve defined on an interval property containing $(t_-(p), t_+(p))$ and satisfying* (b) *and* (c) *above.*

4.42. Proposition: *If $u, t, u + t \in (t_-(p), t_+(p))$, then $\sigma(u + t, p) = \sigma(u, \sigma(t, p))$.*

((d) of Proposition 4.41 applies here.)

4.43. Proposition: *The mappings $p \to t_+(p)$ and $p \to t_-(p)$ are lower and upper semicontinuous respectively.*

(Follows from the proof of 4.40.)

4.44. Definition: *Given a vector field v on M of class C^1, the mapping σ will be called the* **flow** *of v.*

The flow σ is a function of two variables: $p \in M$, $t \in (t_-(p), t_+(p))$. By 4.43, this set of pairs (p, t) is an open subset of $M \times R$, hence a smooth manifold modelled on $E \oplus R$. Let Ω_v denote this manifold. From the proof of Proposition 4.40, we obtain:

4.45. Proposition: *For every v of class C^∞, the flow $\sigma : \Omega_v \to M$ is a smooth mapping.*

Finally we have:

4.46. Proposition: *If $p \in M$, the mapping $\sigma(.\,, p) : (t_-(p), t_+(p)) \to M$ cannot be continuously prolonged to either endpoint, nor can $\sigma(t, p)$ have a limit point as $t \to t_+(p)$ or $t \to t_-(p)$.*

For suppose σ has a limit point as $t \to t_+(p)$. Let p_∞ be this limit point $\sigma(t_+(p), p)$. By the definition of σ, there exists a neighborhood U of p_∞ and an $\varepsilon > 0$ such that σ is defined onto $U \times (-\varepsilon, +\varepsilon)$. Let t_0 be a real number such that $t_+(p) - \frac{1}{2}\varepsilon < t_0 < t_+(p)$ and such that $\sigma(t_0, p) \in U$. Such a point exists. Then define a curve p by $p(t) = \sigma(t, p)$ if $t_- < t < t_+$, $p(t) = \sigma(t - t_0, \sigma(t_0, p))$ if $t_+ \leqq t < t_0 + \varepsilon$. Clearly (from 4.42) $p(t)$ is well defined for $t_- < t < t_+ + \frac{1}{2}\varepsilon$, is smooth and satisfies the differential equation. This contradicts the maximality of $\sigma(t, p)$, as it is described in 4.41(d).

J. Submanifolds

4.47. Definition: *Let M be a smooth E manifold. A subspace $N \subset M$ is called a regularly imbedded submanifold iff there exist a covering of M by domains of charts (U_i = domain of ϕ_i) and a closed linear subspace F of E such that, for every i,*

$$\phi_i(N \cap U_i) = \phi_i(U_i) \cap F.$$

The covering $\{U_i \cap N\}$ and the restrictions $\phi_i | U_i \cap N$ provide a smooth F-structure for N. It is easy to see that this structure does not depend on the particular choice of the original covering $\{U_i\}$ of M.

The following statement can be easily proved.

4.48. Proposition: *Every regularly imbedded submanifold of M is a closed subset of M.*

Examples

1. M and every point in M are regularly imbedded submanifolds.
2. If M is an open set of the Banach space E, then for every closed linear subspace F of E the set $F \cap M$ is a regularly imbedded submanifold of M.

The following proposition provides less trivial examples.

4.49. Proposition: *Let M be a smooth E-manifold and $f\colon M \to \Gamma$ a smooth mapping. If $c \in \mathscr{R}$ is not a critical level for f, the set $N = f^{-1}(\{c\})$ is a regularly imbedded submanifold of M (modelled on a hyperplane of E).*

The reader will see that the lemma below implies the above proposition.

4.50. Lemma: *If $x \in M$, f is defined near x, smooth, and non-horizontal at x, then there exists a chart $y \to \psi(y)$ around x such that $f(y) = \bar{x}(\psi(y)) + c$, where $\bar{x} \in E'$.*

Proof: Considering $f - f(x)$ we may suppose that $f(x) = 0$. Choose a chart ϕ around x such that $\phi(x) = 0$. We know that $x' = (f\phi^{-1})'(0)$ does not vanish. Let $e \in E$ be a vector such that $x'(e) = 1$, and let F be the kernel of x'. Clearly $E = F \oplus |R\, e$. Define a mapping θ by:

$$\theta\,(y \oplus te) = y \oplus [f\phi^{-1}(y \oplus te)]\, e$$

($y \in F$, y and t small).

The derivative at the origin of θ is

$$\theta'(0)\,(y \oplus te) = y \oplus x'(y \oplus te)\, e = y \oplus e,$$

or $\theta'(0)$ = identity. Therefore near the origin θ is a smooth mapping with smooth inverse.

Consider the chart $\psi = \theta\phi$. We have

$$(1) \qquad f\psi^{-1}(y \oplus te) = f\phi^{-1}\theta^{-1}(y \oplus te) = f\phi^{-1}(y \oplus ue)$$

where $\theta(y \oplus ue) = y \oplus te$. But by the definition of θ, this last equality implies that $t = f\phi^{-1}(y \oplus ue)$. From (1) we conclude that

$$f\psi^{-1}(y \oplus te) = t = x'(y \oplus te)$$

Q.E.D.

4.51. More Examples

(a) Let $\mathscr{H}$ be a Hilbert space, $f: \mathscr{H} \to R$ the mapping $f(x) = (x, x)$. f is a quadratic form. Its derivative is $f'(x)\,y = 2\,(x, y)$, and is horizontal only at $x = 0$. Therefore the unit sphere $S = \{x \mid f(x) = 1\}$ is a regularly imbedded submanifold of $\mathscr{H}$ (modelled on any hyperplane).

(b) Let (Ω, Σ, μ) be a measure space, and consider the space $X = L_p(\Omega, \Sigma, \mu)$, $p > 1$. The mapping $f: X \to \mathscr{R}$ defined by

$$f(x) = \int |x(s)|^p\,\mu\,(ds)$$

has as many continuous derivatives as the integer part n of p ("greatest integer less than or equal to p").

Considering X as a manifold of class C^n, a form of Proposition 4.49 applies and we conclude that the unit sphere $S = \{x \mid |x| = 1\} = \{x \mid f(x) = 1\}$ is a regularly imbedded submanifold of X, of class C^n.

(c) Consider again the case of a Hilbert space $\mathscr{H}$, and let S be the unit sphere of the Hilbert space $\mathscr{H} \oplus \mathscr{R}$. According to example (1), S is a smooth manifold modelled on any hyperplane of $\mathscr{H} \oplus \mathscr{R}$, in particular on $\mathscr{H}$. Define now on S the equivalence relation $x \sim y$ iff $x = -y$. The quotient $S/\sim$ has a natural structure as a smooth manifold modelled on $\mathscr{H}$, called the *projective space* on $\mathscr{H}$ and denoted $P(\mathscr{H})$.

K. Riemannian Manifolds

Some Preliminary Remarks on Bilinear Forms

If E is a real Hilbert space we have denoted by $B\,(E, E)$ the linear topological space of bilinear continuous forms on E (see 4.15). The following is clear.

4.52. Lemma: The topology of $B(E, E)$ may be defined by means of the norm

$$(1) \qquad |p| = \sup \{p(x, y);\ |x| \leqq 1,\ |y| \leqq 1\}.$$

Consider now the L.T.S. End (E) as defined in 4.14. Clearly we have

4.53. Lemma: The topology of End (E) may be defined by means of the norm

$$(2) \qquad |u| = \sup \{|u(x)|;\ |x| \leqq 1\}.$$

Our next step is to prove

4.54. Lemma: $B(E, E)$ and End (E) are *isometrically* isomorphic in a canonical way. Symmetrical bilinear forms go onto self-adjoint operators.

Proof: Let $p \in B(E, E)$ and for every $x \in E$, consider the map $x^*: y \to p(x, y)$. Clearly x^* is a continuous linear functional on E. Hence, there exists an element $x^{**} \in E$ such that $x^*(y) = (x^{**}, y)$. Call u_p the map $u_p: x \to x^{**}$. It satisfies

$$(u_p x, y) = p(x, y).$$

Of course $u_p \in$ End (E) and we have a mapping $p \to u_p$ from $B(E, E)$ into End (E).

It is obvious that $p \to u_p$ is linear and onto. Let us note that it is an isometry:

$$|u_p|^2 = \sup_{|x| \leqq 1} |u_p x|^2 = \sup_{\substack{|x| \geqq 1 \\ |y| \leqq 1}} |(u_p x, u_p y)| \leqq |u_p| \sup_{\substack{|x| \leqq 1 \\ |y| \leqq 1}} |(u_p x, y)|$$

$$= |u_p| \sup_{\substack{|x| \leqq 1 \\ |y| \leqq 1}} |p(x, y)| \leqq |u_p|\, |p|,$$

whence $|u_p| \leqq |p|$. Conversely

$$|p| = \sup_{\substack{|x| \leqq 1 \\ |y| \leqq 1}} p(x, y) = \sup_{\substack{|x| \leqq 1 \\ |y| \leqq 1}} (u_p x, y) \leqq |u_p|.$$

Hence $|p| = |u_p|$.

Q.E.D.

4.55. Corollary: $B(E, E)$ is a Frechet space.

Let us consider now two Hilbert spaces F and E (let $(\ ,\ )_f$ and $(\ ,\ )_e$ be their inner products respectively) and assume that $T: F \to E$ is a continuous linear operator. Denote by p the bilinear form on F defined by $p(x, y) = (Tx, Ty)_e$.

4.56. Lemma: The norm of p in $B(F, F)$ is the square of that of T as a bounded operator from F into E.

Proof: By formula (1) we have

$$|p| = \sup \{p(x, y), |x|_f \leqq 1, |y|_f \leqq 1\}.$$

But then

$$|p| = \sup \{(Tx, Ty)_e ; |x|_f \leqq 1, |y|_f \leqq 1\} \leqq |T|^2.$$

On the other hand,

$$|T|^2 = \sup_{|x|_f \leqq 1} (Tx, Tx)_e \leqq \sup_{\substack{|x|_f \leqq 1 \\ |y|_f \leqq 1}} (Tx, Ty) = |p|,$$

and we are through.

Remark: The same result follows from more general statements upon noticing that ${}^tTT = u_p$.

The Definition of Riemannian Manifolds and Some Properties

Let M be a smooth manifold modelled on a Hilbert space E. Since $B(E, E)$ is a Frechet space (Corollary 4.55), Proposition 4.21 allows us to apply Proposition 4.32 and its Corollary 4.33 to conclude that the *duals* of the bundles $T(M)$ and $B(T(M), T(M))$ (notation of 4.32) both exist.

4.57. Definition: The bundles $B(T(M), T(M))$ and $(T(M))'$ will be denoted by $T_2(M)$ and $T_1(M)$ respectively.

4.58. Definition: A pseudometric on M is any smooth symmetric section of $T_2(M)$.

In this definition, the word *symmetric* means that if s is the section, then for every $x \in M$ $s(x) \in B(TM_x, TM_x)$ is symmetric.

4.59. Definition: A pseudometric s will be called a *metric* iff for every $x \in M$, the bilinear form $s(x)$ is an inner product defining a Hilbert space structure on TM_x compatible with its topology.

Of course the set of all pseudo-metrics on M is a linear space; that of the metrics is a cone in it.

4.60. Definition: A *Riemannian manifold* is a pair (M, g) where M is a Hilbert manifold and g is a metric on M.

Assume that g is a pseudo-metric for M. For every element $u \in TM_x$ $(x \in M)$

we shall denote by $|u|$ (or $|u|_g$), the *norm* of u: $|u| = (g(x)(u, u))^{1/2}$. This norm is then a function from $T(M)$ into $\mathscr{R}$.

Assume now that ϕ is a chart of domain $U \subset M$.

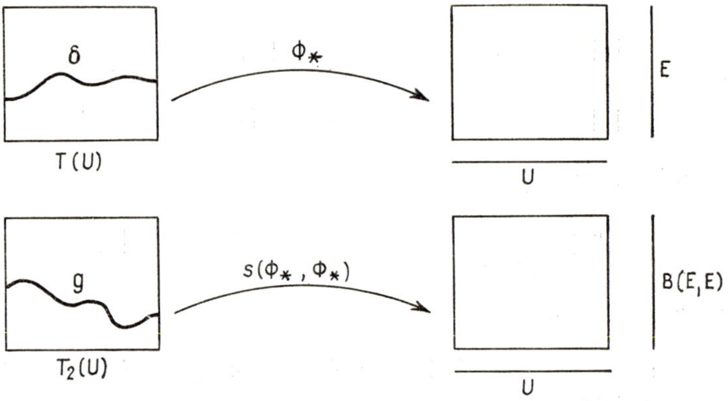

If δ and g are smooth sections of $T(U)$ and $T_2(U)$ respectively, then

$$\begin{aligned} |\delta(x)|_g^2 &= g(x)(\delta(x), \delta(x)) \\ &= g(x)([\phi_*(x)]^{-1}\phi_*(x)\,\delta(x), [\phi_*(x)]^{-1}\phi_*(x)\,\delta(x)) \\ &= [g(x) \circ ([\phi_*(x)]^{-1} \times [\phi_*(x)]^{-1})](\phi_*(x)\,\delta(x), \phi_*(x)\,\delta(x)) \end{aligned}$$

being the composition of $\phi_*(x)\,\delta(x)$ and $g(x) \circ ([\phi_*(x)]^{-1} \times [\phi_*(x)]^{-1})$ is also *smooth.*

This proves that given a pseudometric g, for every smooth vector field δ, the mapping $x \to |\delta(x)|_g^2$ is also smooth.

Assume now that g is a *metric.* Since g induces the inner product $g(x)$ on every TM_x and $\phi_*(x): TM_x \to E$ is continuous it is natural (and useful) to consider the norms $|\phi_*(x)|$ and $|(\phi_*(x))^{-1}|$ of $\phi_*(x)$ and its inverse as operators from one Hilbert space (TM_x with $g(x)$) into another, namly E. First of all, let us define $h: U \to B(E, E)$ by

$$h(x) = g(x) \circ ([\phi_*(x)]^{-1} \times [\phi_*(x)]^{-1}).$$

By definition, h is *smooth.* Moreover, h never assumes on U the value $0 \in B(E, E)$, because g is a metric. Then then the *real function* $x \to |h(x)|$ *is continuous.* Lemma 4.56 implies now that $|h(x)| = |(\phi_*(x))|^{-1/2}$ for every $x \in U$. Hence:

4.61. Proposition: $x \to |(\phi_*(x))^{-1}|$ is continuous on U and never vanishes.

Length of a curve.

Assume that σ is a curve in M, its domain being $[a, b] \subset R$.

4.62. Definition: We define the *length* of σ, $L(\sigma)$, by

$$L(\sigma) = \int_a^b \|p'(t)\|_g \, dt,$$

where $p'(t) = (p_*(t))\,(1)$.

Observe that $L(\sigma)$ might be $+\infty$.

This mapping L from the set of curves in M into $R \cup \{\infty\}$ will be thoroughly studied later; indeed much of our further study is devoted to statements about L (mainly about its critical points, namely, the geodesics). Assume now that M is connected. As in the classical case, topological and arcwise connectedness are equivalent. This follows from the fact that every topologically connected locally arcwise connected space is also arcwise connected. Then for every pair x, y of points in M, there exist smooth curves p joining them: $p(a) = x$, $p(b) = y$. Consider all the numbers $L(p)$.

4.63. Definition: $d(x, y) = \inf L(p)$, the infininum taken on the (non-void) set of smooth curves joining x and y.

4.64. Proposition: The mapping d is a distance on M defining the original topology.

Proof: It suffices to prove that d induces the original topology on some open neighborhood of every point. Assume then that ϕ is a chart and U an open set in the domain of ϕ on which $|(\phi_*(x))^{-1}|$ is bounded (since $|\phi_*(x))^{-1}|$ is continuous by 4.61, such a U does exist) and such that $\phi(U)$ is a ball in E. Such U's cover M and are d-open. We prove now that d induces the relative topology on U. If $y, z \in U$, let

$$p(t) = \phi^{-1}(t\phi(y) + (1-t)\,\phi(z)), \quad 0 \leqq t \leqq 1.$$

Then

$$\phi_*(p(t))\,p'(t) = \frac{d}{du}\phi\,(p(u))_{u=t} = \phi(y) - \phi(z)$$

and hence

$$p'(t) = [\phi_*(p(t))]^{-1}\,(\phi(y) - \phi(z)).$$

Therefore

$$|p'(t)| \leqq |(\phi_*(p(t)))^{-1}|\,|\phi(y) - \phi(z)|.$$

Since we have assumed that $|(\phi_*(x))^{-1}|$ is bounded, we have

$$|p'(t)| \leqq K\,|\phi(y) - \phi(z)|,$$

from which it follows that

$$L(p) \leqq K\,|\phi(y) - \phi(z)|$$

Finally:

$$d(y, z) \leqq K \int_0^1 |t\phi(y) + (1 - t)\phi(z)|\, dt.$$

This implies that if $z \to y$ in the topology of M, then $d(y, z) \to 0$. On the other hand, if q is any curve joining y and z, we have

$$L(q) = \int_0^1 |q'(t)|\, dt \geqq \frac{1}{K}\int_0^1 |\phi_*(q(t))\, q'(t)|\, dt \geqq \frac{1}{K}\left|\int_0^1 \phi_*(q(t))\, q'(t)\, dt\right|$$

$$= \frac{1}{K}\left|\int_0^1 \frac{d}{dt}(\phi q(t))\, dt\right| = \frac{1}{K}|\phi(y) - \phi(z)|,$$

and then, choosing $L(q)$ near $d(y, z)$, we see that when $d(z, y) \to 0$ it necessarily follows that $|\phi(y) - \phi(z)| \to 0$, which implies that $z \to y$ in M.
Q.E.D.

Remark: Of course even if M is not connected it is possible to define a distance on M induced by the distance defined above on each of the components.

4.65. Definition: A Riemannian manifold is said to be *complete* iff each of its components is complete under the metric defined in 4.63.

Generally, we shall deal with complete Riemannian manifolds.

Gradient

Let (M, g) be a Riemannian manifold. Every TM_x is endowed with an inner product $g(x)$ that makes it into a Hilbert space. That means that there exist canonical isometries $\tau_x : (TM_x)' \to TM_x$.

4.66. Definition: If f is a smooth function on M, the *gradient* of f is the vector field ∇f defined by $(\nabla f)_x = \tau_x f_*(x)$. That is ∇f satisfies

$$g(x)((\nabla f)_x, v) = vf,$$

for all $v \in TM_x$ and all $x \in M$.

It is clear that $f \to \nabla f$ is a linear mapping from $C^\infty(M)$ into $\Omega(M)$. Moreover, if $\nabla f \equiv 0$, then f is constant on every component of M. Hence, if M is connected, ∇ is an injection of $C^\infty(M)/R$ into $\Omega(M)$.

Part 2

Morse Theory

A. The Non-critical Neck Principle

Assume that M is a smooth manifold modelled on a Banach space E. We recall that, given a smooth function $f\colon M \to R$, a real number c is called a *critical level* for f iff there exists a point $x \in M$ such that $f(x) = c$ and $f_*(x) = 0$ (see 4.4 and the notation in 4.9).

4.67. Theorem (Non-critical Neck Principle): Let f be a smooth function $f\colon M \to \mathscr{R}$ and consider the sets $N \subset W \subset M_1 \subset M$ defined by:

$$W = \{x \in M;\ a < f(x) < b\}$$

$$M_1 = \{x \in M;\ a - \varepsilon < f(x) < b + \varepsilon\}$$

$$N = f^{-1}(c),$$

where a, b, c and ε are real numbers, $\varepsilon > 0$ and $a < c < b$. Suppose that v is a smooth vector field defined on M, and let $\sigma(t, p)$ denote its flow (see 4.44). Assume also that

(a) $$vf \geqq \delta > 0$$

and that

(b) for every fixed $p \in N$, the function $f(\sigma(t, p))$ assumes values greater than b and less than a in its interval of definition $t_-(p) < t < t_+(p)$.

Then the manifolds W and $(a, b) \times N$ are diffeomorphic.

Remark: The fact that N is a manifold follows from 4.49.

Proof: Without loss of generality we may assume that $a = -1$, $b = +1$, $c = 0$ and $vf = 1$ (this last condition is achieved by replacing the original v by $(vf)^{-1}v$). That means that

$$\frac{d}{dt} f(\sigma(t, p)) = vf = 1,$$

and hence (after integrating), that:

$$[*] \qquad f(\sigma(t, p)) = f(p) + t.$$

Assume now that $n \in N$. Then $f(\sigma(t, n)) = t$ and, from (b), we conclude that $t_-(n) < -1$, $t_+(n) > +1$. This means in particular that the domain of the mapping σ contains $(-1, +1) \times N$. Consider now the restriction

$$\Lambda : (-1, +1) \times N \to M$$

of σ (i.e., $\Lambda(t, n) = \sigma(t, n)$).

We shall show that Λ is a diffeomorphism between $(-1, +1) \times N$ and W.

1. *Λ assumes values in W.* In fact, from [*] we obtain $|f(\Lambda(t, n))| = |f(\sigma(t, n))| = |t| < 1$.

2. *Λ maps $(-1, +1) \times N$ onto W.* Assume that $p \in W$. Set $t = f(p)$, $n = \sigma(-t, p) = \sigma(-f(p), p)$.

Then $f(n) = f(\sigma(-f(p), p))$, and by [*], we have $f(n) = -f(p) + f(p) = 0$, whence $n \in N$. Clearly (see the definition of W) we also have $-1 < t < +1$. Now using Proposition 4.42 we obtain:

$$\Lambda(t, n) = \sigma(t, n) = \sigma(f(p), \sigma(-f(p), p)) = \sigma(0, p) = p;$$

and Λ is therefore onto.

3. *Λ is smooth and has a smooth inverse.* If $p = \Lambda(t, n)$, using 4.42 again and also using [*] we conclude that

$$[**] \qquad \begin{aligned} t &= f(p) \\ n &= \sigma(-t, p) = \sigma(-f(p), p). \end{aligned}$$

But then 4.45 implies that both Λ and Λ^{-1} are smooth.

Remark 1: If the hypothesis (b) is not satisfied then the proposition is false, as the following example shows: let V be the surface of a vertical circular cylinder in Euclidean space R^3. Denote by $f(x)$ the vertical coordinate of a point $x \in V$ and by $v(x)$ a vertical unit vector with origin at $x \in V$. Now remove a point $z \in V$ such that $f(z) = 0$ and let M denote the remaining manifold.

For the values $a = -1$, $b = +1$ the proposition does not hold for M. Nevertheless, all the hypotheses except (b) are satisfied.

Remark 2: Observe that the diffeomorphism $\Lambda : (a, b) \times N \to W$ sends $\{z\} \times N$ diffeomorphically onto $f^{-1}(z)$ for every $a < z < b$ (this follows from the formula [**] in the above proof).

4.68. Proposition: Assume the same hypothesis as in 4.67 and furthermore assume that (b) is satisfied in the following stronger version:

(b′) for every fixed $p \in N$, the function $f(\sigma(t, n))$ assumes values greater than $b + \eta$ and less than $a - \eta$ for some $\eta > 0$ (the same for all $p \in N$).

Then if $W_1 = \{x \in M; a \leqq f(x) \leqq b\}$, there exists a diffeomorphism between W_1 and $N \times [a, b]$ (as long as they are manifolds, i.e. when N is a manifold without boundary).

Proof: From the proposition above we obtain the existence of a mapping

$$\Lambda : N \times \left(a - \frac{\eta}{2}, b + \frac{\eta}{2}\right) \to W_2 \tag{1}$$

where $W_2 = \{x \in M; a - \eta/2 < f(x) < b + \eta/2\}$.

This mapping sends $N \times \{z\}$ onto $f^{-1}(z)$ for every $a - \eta/2 < z < b + \eta/2$. Then the restriction of Λ to $N \times [a, b]$ is the desired diffeomorphism.

4.69. Corollary: Under the hypothesis of 4.68, there exists a homotopy $H: M \times I \to M$, where $I = [0, 1]$, such that if $H_s(m) = H(m, s)$ then

1. for every $s \in I$, $H_s: M \to M$ is a diffeomorphism;
2. if $m \in M$ does not satisfy $a - \eta/4 \leqq f(m) \leqq b + \eta/4$, then $H_s(m) = m$ for all s;
3. $H_0 =$ identity;
4. $H_1(\{x; f(x) \leqq a\}) = \{x; f(x) \leqq b\}$.

Proof: Let h be a smooth function as shown below such that $h'(x) > 0$ for all x.

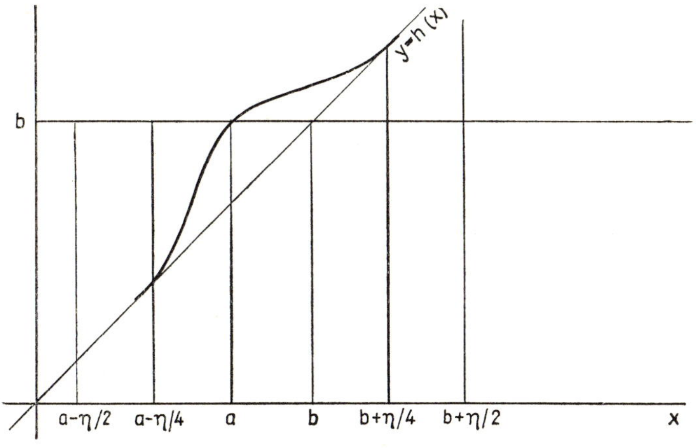

Let $F = \{x; a - \eta/4 \leqq f(x) \leqq b + \eta/4\}$ and let $G = M - F$. Clearly G is open and $M = G \cup W_2$.

Now if Λ is the mapping defined in (1) of the proof of 4.68, then for every s between 0 and 1 we define H_s as follows:

$$\text{if} \quad m \in G, \quad \text{then} \quad H_s(m) = m$$

$$\text{if} \quad m = \Lambda(n, t) \in W_2, \quad \text{then} \quad H_s(m) = \Lambda(n, (1 - s)t + sh(t)).$$

Observe that H_s is well defined (and equal to the identity) on $W_2 \cap G$.

Indeed, the reader can now verify that the H_s have the properties 1), ..., 4).

Remark: Part (4) of this corollary says in particular that $\{x; f(x) \leqq a\}$ and $\{x; f(x) \leqq b\}$ are diffeomorphic. This fact will be used very often. An important generalization appears in: 4.72.

B. The Palais-Smale Condition

Let us assume now that (M, g) is a Riemannian manifold.

4.70. Definition: If $f \in C^\infty(M)$, we shall that f satisfies the *Palais-Smale condition* ("P-S condition") iff whenever S is a set in M on which f is bounded and $\|\nabla f\|$ is not bounded away from zero, then there exists a critical point of f adherent to S.

Of course this is equivalent to: if $\{x_n\}$ is a sequence in M such that $f(x_n)$ is bounded and $\|(\nabla f)_{x_n}\| \to 0$, then there exists a *convergent* subsequence of $\{x_n\}$ (the limit being necessarily a critical point).

Remark: This condition appears in (2) and (3) of the bibliography.

4.71. Theorem (non-critical neck principle for Riemannian manifolds): Let M be a complete Riemannian manifold, $f \in C^\infty(M)$ and consider the sets $N \subset W \subset W_1 \subset M_1 \subset M$ defined by

$$W = \{x; a < f(x) < b\}$$

$$W_1 = \{x; a \leqq f(x) \leqq b\}$$

$$M_1 = \{x; a - \varepsilon < f(x) < b + \varepsilon\}$$

$$N = f^{-1}(c),$$

where a, b, c and ε are real numbers, $\varepsilon > 0$ and $a < c < b$. Assume that the following hypotheses are satisfied:

(α) f satisfies the P-S condition on M_1;
(β) f has no critical point in M_1.

Then W is diffeomorphic to $N \times (a, b)$ and (if N has empty boundary) W_1 is diffeomorphic to $N \times [a, b]$. Moreover, the diffeomorphism may be chosen so as to send $N \times \{z\}$ onto $f^{-1}(z)$ diffeomorphically for every z between a and b.

Proof: Define a vector field v by $v = \nabla f$ (and let σ be its flow). We shall show that v and f satisfy hypothesis (a) of 4.67 and (b′) of 4.68.

First of all, we see that $vf = (\nabla f) = (\nabla f, \nabla f) = \|\nabla f\|^2$ and that hypotheses (α) and (β) imply that $\|\nabla f\|$ is bounded away from 0 on $M_2 = \{x;\ a - \varepsilon < f(x) < b + \varepsilon/2\}$, i.e., $\|\nabla f(z)\| \geqq \delta > 0$ if $t \in M_2$. Thus, hypothesis (a) of 4.67 is satisfied. So is hypothesis (b′). In fact, assume that

[*] $$f(\sigma(t, p)) \leqq b + \frac{\varepsilon}{2} \quad \text{for} \quad 0 \leqq t < t_+ = t_+(p).$$

Then

$$\frac{d}{dt} f(\sigma(t, p)) = \frac{d\sigma}{dt} f = (\nabla f) f = \|\nabla f(\sigma(t, p))\|^2,$$

and hence

[**] $$\int_0^{t_+} \|\nabla f(\sigma(t, p))\|^2\, dt = \lim_{u \to t_+} \int_0^u \frac{d}{dt} f(\sigma(t, p))\, dt$$
$$\leqq \sup_{0 \leqq u \leqq t_+} f(\sigma(u, p)) - f(p) \leqq b + \frac{\varepsilon}{2} - f(p).$$

Since $\|\nabla f\| \geqq \delta$ on M_2, it follows from [*] and [**] that

$$\delta t_* \leqq b + \frac{\varepsilon}{2} - f(p)$$

whence t_+ is finite.

But then we also have (from [**]):

[***] $$\int_0^{t_+} \|\nabla f(\sigma(t, p))\|\, dt < +\infty.$$

Since σ is the flow of ∇f, we conclude that $\nabla f(\sigma(t, p)) = \dfrac{d\sigma}{dt}(t, p)$, and then [***] implies:

[****] $$\int_0^{t_+} \left\| \frac{d\sigma}{dt}(t, p) \right\| dt < +\infty.$$

Assume now that $\eta > 0$ is given. Then $T < t_+$ may be chosen so that

$$\int_T^{t_+} \left\| \frac{d\sigma}{dt}(t, p) \right\| dt < \eta.$$

Consider now two points $\sigma(x, p)$, $\sigma(y, p)$ with $T \leqq x \leqq y < t_+$. They may be joined by a curve $\gamma(t) = \sigma(t, p)$, $x \leqq t \leqq y$, whose length by the last formula is less than η. Therefore the distance $\varrho(x, y) < \eta$ and we have proved thereby that the net $\sigma(p, t)$, $0 \leqq t < t_+$ is a *Cauchy net* (cf. Kelly, *General Topology*). Since M is complete, $\lim_{t \to t_+} \sigma(t, p)$ exists, which together with $t_+ < +\infty$, $t \to t_+$, contradicts Prop. 4.45. Thus hypothesis (b′) of 4.68 is satisfied (for $\eta = \varepsilon/2$) and then 4.68 applies.

Q.E.D.

Remark: Observe that hypotheses (α) and (β) are independent of c. Therefore, the conclusion is true for any c between a and b.

4.72. Corollary: Under the hypotheses of Theorem 4.71, there exists a homotopy $H: M \times I \to M$ ($I = [0, 1]$) having the properties:

1. for every $s \in I$, $H_s: M \to M$ is a diffeomorphism;
2. if $m \in M$ does not satisfy $a - \varepsilon/8 \leqq f(m) \leqq b + \varepsilon/8$, then $H_s(m) = m$ for all s;
3. $H_0 =$ identity;
4. $H_1(\{x; f(x) \leqq a\}) = \{x; f(x) \leqq b\}$.

Proof: We have shown that the hypotheses of 4.71 imply those of 4.68, and hence of its corollary.

C. Local Study of Critical Points

Let E be a Hilbert space and f a smooth real function defined on a neighborhood of $0 \in E$. Using Taylor's expansion we write

$$f(x) = f(0) + f'(0)(x) + \tfrac{1}{2} f''(0)(x, x) + R(x),$$

where $R(x)$ is a function of order 3.

Assume that 0 is a critical point of f. Then:

$$[*] \qquad f(x) = f(0) + \tfrac{1}{2} f''(0)(x, x) + R(x).$$

Since the bilinear form $f''(0)$ is continuous and symmetric, there exists (see 4.54) a (unique) symmetric operator $A \in \operatorname{End}(E)$ such that

$$f''(0)(x, y) = (Ax, y) = (x, Ay).$$

Formula [*] then becomes

$$[**] \qquad f(x) = f(0) + \tfrac{1}{2}(Ax, x) + R(x).$$

Let us consider a smooth change of coordinates $y \to x(y)$, such that $x(0) = 0$. Then $f(y) = f_1(x(y))$ and using the chain rule we obtain:

$$f''(y)(z_1, z_2) = f_1''(x(y))(x'(y)z_1, x'(y)z_2) + f'(x(y))(x''(y)(z_1, z_2)).$$

Since 0 is a *critical point*, we get:

$$f''(0)(z_1, z_2) = f_1''(0)(x'(0)z_1, x'(0)z_2).$$

This formula shows that the operator A transforms according to:

$$[***] \qquad A = u^{-1}A_1u, \quad \text{where} \quad u = x'(0).$$

4.73. Definition: Let f be a smooth function defined on some Riemannian manifold, x a critical point of f. We shall say that x is a *non-degenerate critical point of* f iff in any chart, the operator A defined in [**] is invertible.

Remark: The formula [***] shows this notion to be coordinate independent. We are led to the same concept as follows. *If x is a critical point* and ϕ is a chart around x, the bilinear form $H(f)_x$ defined on TM_x by

$$H(f)_x(\mu, \lambda) = [(f\phi^{-1})''(\phi(x))](\phi_*(x)\mu, \phi_*(x)\lambda)$$

does not depend on ϕ. Hence we may make the following definition.

4.74. Definition: The bilinear form $H(f)_x$ is called the *Hessian of f at x.*

According to our definitions, the Hessian of a smooth function is a smooth section of the bundle $B(T(M), T(M)) = T_2(M)$ defined on the set of critical points of f.

4.75. Definition: A critical point x is called *non-degenerate* if $H(f)_x$ is a scalar product defining the given topology of TM_x.

It is obvious that 4.73 and 4.75 are equivalent.

Remark: A very elegant definition of $H(f)_x$ is given in Milnor, "Morse Theory" (Ann. of Math. Studies, No. 51, Princeton 1963).

Let F be a complex Hilbert space. Denote by End (F) the space of all the continuous linear operators $T: F \to F$ and by Aut (F) (respectively $H(F)$) the subset of invertible operators (respectively, the subspace of the Hermitean operators.

4.76. Lemma: If $A \in \text{Aut}(F) \cap H(F)$, then the mapping $\psi: \text{End}(F) \to H(F) \times H(F)$ defined by

$$\psi(B) = (B^*A + AB, i(B^*A - AB))$$

is one-to-one onto and both ψ and ψ^{-1} are continuous.

Proof: In fact, given S and T symmetric, define

$$B = \tfrac{1}{2}A^{-1}(S + iT).$$

Then $B^*A + AB = S$ and $i(B^*A - AB) = T$, and hence

$$(S, T) \to \tfrac{1}{2}A^{-1}(S + iT)$$

is the inverse of ψ. Clearly both are continuous.

4.77. Lemma: Assume that $A \in \text{Aut}(F) \cap H(F)$. Then the mapping $\phi: \text{Aut}(F) \to H(F) \times H(F)$ defined by

$$\phi(B) = (B^*AB, i(B^*AB^{-1} - A))$$

is differentiable and its derivative at $B = I$ is

$$\delta\phi(I, B) = \psi(B).$$

Proof: Observe that $\delta(B^{-1}) = -\delta B$ and compute.

4.78. Corollary: ϕ maps a neighborhood of $I \in \text{Aut}(F)$ diffeomorphically onto a neighborhood of $(A, 0) = \phi(I)$.

Proof: Use the implicit function theorem.

Assume now that $x \to (A(x), D(x))$ is a smooth mapping from an open set of F into $H(F) \times H(F)$ such that $A(0) = A$ and $D(0) = 0$. Then

4.79. Lemma: $x \to B(x) = \phi^{-1}(A(x), D(x))$ is a smooth mapping such that $B(0) = I$ and $A(x) = B^*(x)\, A(0)\, B(x)$ and

$$D(x) = i(B^*(x)\, A(0)\, B^{-1}(x) - A(0)).$$

Proof: Follows from the corollary above.

Assume now that E is a *real* Hilbert space.

4.80. Proposition: Let $x \to A(x)$ be a smooth mapping from a neighborhood of $0 \in E$ into End (E) such that $A(x)$ is symmetric and $A(0)$ is also invertible. Then there exists a smooth mapping $x \to B(x)$ of some neighborhood of $0 \in E$ into Aut (E) such that:

$$A(x) = B^*(x)\, A(0)\, B(x).$$

Proof: Let F be the complexification of $E: F = E \otimes C = E \oplus iE$. Define the mappings $x + iy = z \to \bar{A}(z)$ by $\bar{A}(z)(u + iv) = A(x)u + iv$ and $z \to \bar{D}(z)$ by $\bar{D}(z) = 0$. Apply Lemma 4.79 to prove that there exists a map $z \to \bar{B}(z)$ such that:

(1) $$\bar{B}^*(z)\, \bar{A}(0)\, \bar{B}(z) = \bar{A}(z)$$

(2) $$\bar{B}^*(z)\, \bar{A}(0)\, (\bar{B}(z))^{-1} = \bar{A}(0).$$

From (1) and (2) we conclude that:

(3) $$(\bar{B}(z))^2 = (\bar{A}(0))^{-1}\, \bar{A}(z)$$

(4) $$(\bar{B}^*(z))^2 = \bar{A}^*(z)\, (\bar{A}^*(0))^{-1}.$$

Now observe that for every $x \in E$, $\bar{A}(x)$ and $A(0)$ leave E invariant: $\bar{A}(x)E = A(x)E \subset E$, $\bar{A}(0)E = E$. Then (3) and (4) imply that E is also invariant under $(\bar{B}(x))^2$ and $(\bar{B}^*(x))^2$. Now we shall use the following statement (see below for a justification):

(5) If an operator T satisfies $\|I - T^2\| < 1$, then the invariant (closed) subspaces of T and T^2 are the same.

From (5), it follows that E is also invariant under $B(x)$ and $B^*(x)$, provided that $x \in E$ is near 0. Hence, by restriction to E (and calling $B(x) = \bar{B}(x)|_E$) we obtain from 1

$$B^*(x)\, A(0)\, B(x) = A(x), \quad x \in E, \quad x \text{ near } 0$$

as desired.

Justification of (5): observe that

$$T = (1 - (1 - T^2))^{1/2} = 1 + \frac{1}{2}(1 - T^2) - \frac{1}{8}(1 - T^2)^2$$

$$+ (-1)^{n-1} \frac{(2(n-1))!}{2^{2n-1} \cdot n \cdot ((n-1)!)^2} (1 - T^2)^n + \cdots$$

where the series converges in the uniform topology of operators if $|1 - T|^2 < 1$.

4.81. Proposition: Let A be a symmetric invertible operator in End (E) (E = real Hilbert space). Then there exist $T \in \text{Aut}(E)$ and a projector $P \in \text{End}(E)$ such that

$$(Ax, x) = \|PTx\|^2 - \|(1 - P)Tx\|^2, \quad x \in E.$$

Proof: Let h be the characteristic function of $[0, \infty)$ and g the function $g(\lambda) = |\lambda|^{-1/2}$, λ = real $\neq 0$. Since A is invertible, g is continuous on the spectrum of A. Then $S = g(A)$ is defined.

Clearly S (being a function of A) is symmetric and commutes with A. Moreover S is invertible (because $g \neq 0$ on Spectrum (A)). Call $T = S^{-1}$. T is symmetric and invertible.

Now define $P = h(A)$. P is clearly a projector (because $h^2 = h$) commuting with A, hence also with T. Since we have

$$\lambda\,(g(\lambda))^2 = h(\lambda) - (1 - h(\lambda)),$$

we conclude that $AT^{-2} = P - (1 - P)$, and hence $A = PT^2 - (1 - P)T^2$, But then

$$(Ax, x) = (PT^2x, x) - ((1 - P)\,T^2x, x) = \|PTx\|^2 - \|(1 - P)\,Tx\|^2,$$

as desired.

4.82. Proposition (Morse Lemma): Let f be a smooth function defined on a Riemannian manifold M (modelled on E). If x is a non-degenerate critical point of f, then there exists a chart ϕ around x (sending x into 0) and a projector P in E such that

(1) $$f(y) = f(x) + |P\phi y|^2 - |(1 - P)\phi y|^2$$

when y belongs to the domain of ϕ

Proof: Let ψ be any chart around x (sending x into 0) and put $g(y) = (f\psi^{-1})\,(y) - f(x)$, where $e \in E$ and a near 0. Then

(2) $$g(y) = g(0) + g'(0)\,y + \int_0^1 [g''(sy)]\,(y, y)\,(1 - s)\,ds$$

$$= \left[\int_0^1 (1 - s)\,g''(sy)\,ds\right](y, y)$$

Now, since $\alpha_y = \int_0^1 (1 - s)\,g''(sy)\,ds$ is a symmetric bilinear form on E, there exists a mapping $y \to A(y) \in \text{End}(E)$ defined by

(3) $$(A(y)\,x, y) = \alpha_y\,(x, y).$$

Clearly every $A(y)$ is symmetric and

$$(A(0)\,x, y) = \alpha_0\,(x, y) = \tfrac{1}{2}g''(0)\,(x, y).$$

Together with the fact that x is a *nondegenerate* critical point of f this implies that $A(0)$ is invertible. Now we apply 4.80 and obtain $B(x)$ satisfying:

$$A(x) = B^*(x)\,A(0)\,B(x).$$

This implies that

(4) $$(A(y)\,y, y) = (A(0)\,B(y)\,y, B(y)\,y),$$

and from (2), (3) and (4) we obtain:

$$g(y) = (A(0)\,B(y)\,y, B(y)\,y).$$

Using 4.81, we conclude that there exist T and P such that

(5) $$g(y) = |PT\,B(y)\,y|^2 - |(1-P)\,T\,B(y)\,y|^2.$$

Define ϕ by

$$\phi(y) = T\,(B\,(\psi(y))\,\psi(y)).$$

By (5), we have:

$$f(y) = f(x) + g\,(\psi(y)) = f(x) + |PT\,B\,(\psi(y))\,\psi(y)|^2$$
$$- |(1-P)\,T\,B\,(\psi(y))\,\psi(y)|^2 = f(x) - |P\phi x|^2 - |(1-P)\,\phi x|^2,$$

as desired.

We must observe that ϕ is actually a chart; indeed it is the composition of

$$y \to \psi(y)$$

$$e \to B(e)\,e$$

$$z \to T(z)$$

and clearly the first and third mappings are diffeomorphisms while $y \to B(y)\,y$, having the identity as derivative at the origin (compute!) is also a local diffeomorphism. Hence ϕ is a chart on some domain around x.

Q.E.D.

D. Global Study of Critical Points

We begin by defining *handles* and the *attaching of handles*.

4.83. Definition: For every cardinal number k we shall denote by D^k the closed unit ball around 0 in a Hilbert space having an orthonormal basis of cardinality k. ∂D^k will denote its boundary.

4.84. Definition: Let M and $\bar{M}$ be two smooth manifolds (possibly with boundary).

We shall say that $\bar{M}$ has been obtained from M by *attaching a handle of type* (k, l) if the following conditions are satisfied:

1. M is a regularly embedded submanifold of $\bar{M}$;
2. There exists a closed subset $H \subset \bar{M}$ and a mapping $h: D^k \times D^l \to H$, such that:

2a. $M \cup H = \bar{M}$,

2b. h is a homeomorphism,

2c. $h(\partial D^k \times D^l) = H \cap M \subset \partial M$,

2d. $H - M$ is a submanifold of $\bar{M}$ with boundary,

2e. the restriction $h \big| (\mathring{D}^k \times D^l)$ is a diffeomorphism of $\mathring{D}^k \times D^l$ onto $H - M$, and

2f. the restriction $h | (\partial D^k \times D^l)$ is a regular embedding of $\partial D^k \times D^l$ into ∂M.

In this situation, we use the following notation for $\bar{M}$: $\bar{M} = M \bigcup_h H(k, l)$.

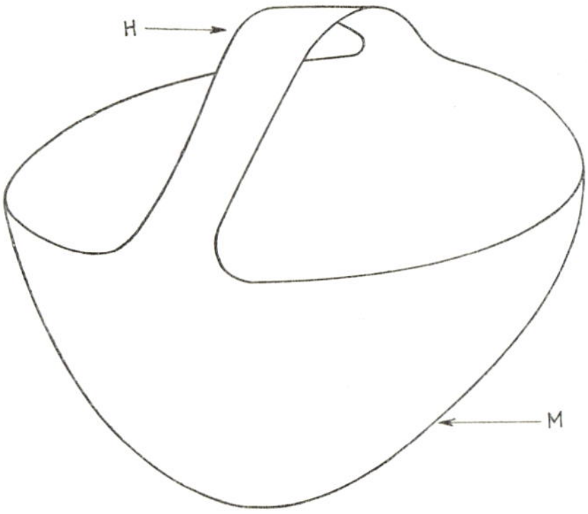

Remark: Obviously $\dim \bar{M} = k + l$.

We shall say that H is a *handle*.

As a generalization, we state the following definition:

4.85. Definition: We shall say that $\bar{M}$ has been obtained from M by attaching n handles of types $(k_1, l_1), \ldots, (k_n, l_n)$ if separately attaching n such handles to M in such a way that $h_i(H_i) \cap h_j(H_j) = \varnothing$ the manifold obtained is $\bar{M}$.

In this situation, we use the following notation for $\bar{M}$: $\bar{M} = M \bigcup_{h_1} H_1 \cdots \bigcup_{h_n} H_n = \bar{M}$. (See next figure.)

Assume now that f is a smooth real-valued function defined on some Riemannian manifold M (modelled on E) and that $x \in M$ is a non-degenerate critical point of f. By 4.82, f may be represented, in some chart ϕ around x, as

$$f(y) = f(x) + |P\phi y|^2 - |(1 - P)\phi y|^2,$$

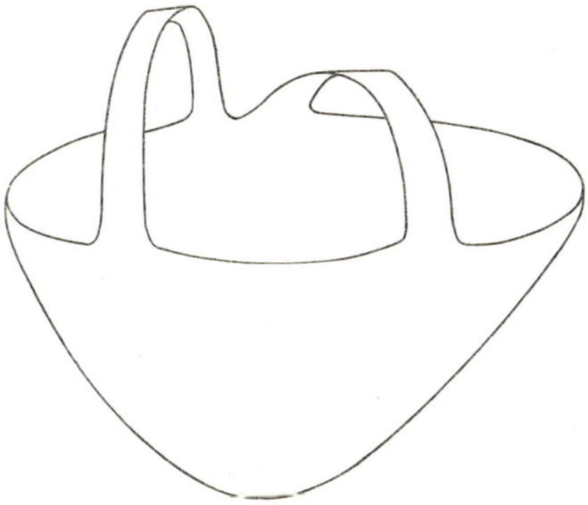

where P is a projector in E.

4.86. Definition: The index of f at x (or the index of x, if there is no confusion) is the pair (k, l), where $k = \dim(PE)$, $l = \dim((1 - P)E)$.

Of course both k and l may be infinite cardinal numbers.

We arrive now at the most important theorems of this section.

The symbol $\sim$ will mean "is diffeomorphic to".

Let (M, g) be a complete Riemannian manifold.

Given $f \in C^\infty(M)$ then for every $s, t \in \mathscr{R}$, define

$$[f \leqq s] = \{x \in M; f(x) \leqq s\}$$

$$[s \leqq f \leqq t] = \{x \in M; s \leqq f(x) \leqq t\}.$$

4.87. Theorem: (Critical neck principle.) Assume that $f \in C^\infty(M)$ and that the only critical level of f between a and b is c, $a < c < b$; a and b are supposed to be non-critical. Assume also that f satisfies the P-S condition on $[a \leqq f \leqq b]$ and that its critical points $p_1, \ldots, p_n$ (at level c) are non-degenerate, their indices being (k_i, l_i), $i = 1, \ldots, n$. Then

$$[*] \qquad [f \leqq b] \sim [f \leqq a] \underset{h_1}{\cup} H_1 \cdots \underset{h_n}{\cup} H_n,$$

where (for every $i = 1, \ldots, n$), H_i is a (k_i, l_i) handle.

Remark: The number of critical points is finite.

Proof: 1. Considering $f-c$ instead of f, we may assume that $c = 0, a<0<b$.

2. Moreover, by 4.72, it follows that

$$[f \leqq s] \sim [f \leqq t] \quad \text{if} \quad a \leqq s, t < 0$$

and

$$[f \leqq s] \sim [f \leqq t] \quad \text{if} \quad 0 < s, t \leqq b.$$

Hence [*] is equivalent to

$$[f \leqq s] \sim [f \leqq t] \underset{h_1}{\cup} H_1 \cdots \underset{h_n}{\cup} H_n$$

for some s, t, with $a \leqq s < 0 < t \leqq b$.

3. Consider now charts $\phi_1, \ldots, \phi_n$ around $p_1, \ldots, p_n$, sending p_i into $0 \in E$ and such that, for x near p_i,

$$f(x) = \|P_i\phi_i x\|^2 - \|Q_i\phi_i x\|^2,$$

where $P_i = 1 - Q_i$ is a projector of E.

The existence of such charts follows from 4.82.

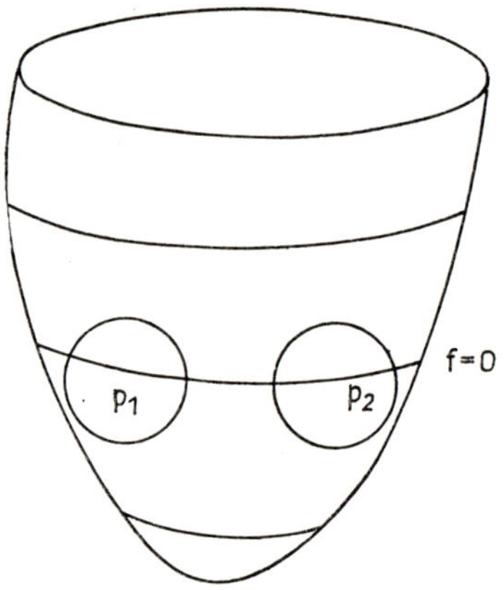

The ball of radius $10r$ is mapped by ϕ_i^{-1} into a neighborhood $U_i(r)$ of p_i and the chart $\phi_i^r: x \to \dfrac{1}{r}\phi_i(x)$ sends $U_i(r)$ onto the ball of radius 10 in E: call this ball B. We have

$$f(x) = r^2(\|P\phi_i^r x\|^2 - \|Q\phi_i^r x\|^2).$$

If r is chosen small enough, we have $ar^{-2} < -1$, and $2 < br^{-2}$. Replacing f by $r^{-2}f$ and ϕ_i by ϕ_i^r, we reduce the problem to the case in which there are charts $\phi_1, \ldots, \phi_n$ around $p_1, \ldots, p_n$ whose domains $U_1, \ldots, U_n$ have disjoint closures and which are all mapped onto the ball of radius 10 around $0 \in E$.

3. Now f may be expressed near p_i as

$$f(x) = \|P_i\phi_i x\|^2 - \|Q_i\phi_i x\|^2,$$

where $P_i = 1 - Q_i$ is a projector of E.

All the levels c between α and β are non-critical (α and β included) except $c = 0$, where α and β satisfy $\alpha < -1, 2 < \beta$.

4. The old sets $[f \leqq a]$ and $[f \leqq b]$ correspond to the new levels α and β and then coincide with the new $[f \leqq \alpha]$ and $[f \leqq \beta]$; but, from 2. we have $[f \leqq \alpha] \sim [f \leqq -1]$ and $[f \leqq 2] \leqq [f \leqq \beta]$ and hence [*] now becomes

(4a) $$[f \leqq 2] \sim [f \leqq -1] \bigcup_{h_1} H_1 \cdots \bigcup_{h_n} H_n.$$

5. To simply notations, define u_i and w_i on U_i by

$$u_i(x) = \|P\phi_i x\|$$

$$v_i(x) = \|Q\phi_i x\|.$$

Then we have

$$f(x) = (u_i(x))^2 - (v_i(x))^2, \quad x \in U_i.$$

6. The remaining proof will depend on the existence of a function Λ having the following properties:

(6a) $\Lambda \in C^\infty(M)$

(6b) $\Lambda \geqq 0; \Lambda = 0$ outside $U_1 \cup \cdots \cup U_n$;

(6c) $\Lambda(p_i) = \frac{3}{2}, i = 1, \ldots, n$;

(6d) if $f(x) \geqq 2$, then $\Lambda(x) = 0$.

Put $g = f - \Lambda$; we also require that

(6h) g has the same critical points as f;

(6j) g satisfies the P-S condition;

(6k) $$[f \leqq -1] \bigcup_{h_1} H_1 \cdots \bigcup_{h_n} H_n = [g \leqq -1].$$

7. Let us first see why the existence of such a Λ implies (4a) and hence the theorem. First of all, from (6b) and (6d) it follows that

(7a) $$[f \leqq 2] = [g \leqq 2].$$

On the other hand, by (6c), $g(p_i) = -\frac{3}{2}$. Hence, by (6h), all the levels between -1 and 2 are non-critical for g.

That means (see 4.72) that

(7b) $$[g \leqq -1] \sim [g \leqq 2].$$

Finally, from (6k), (7a) and (7b) we obtain (4a), as desired.

8. *Existence of Λ*

Let λ and η be two smooth real functions of a real variable as in the diagrams below:

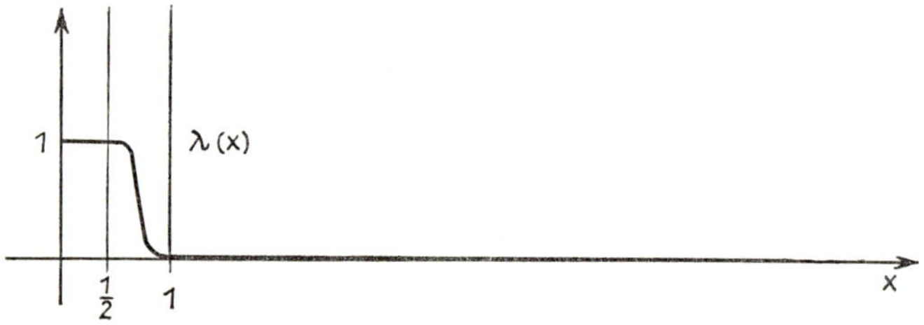

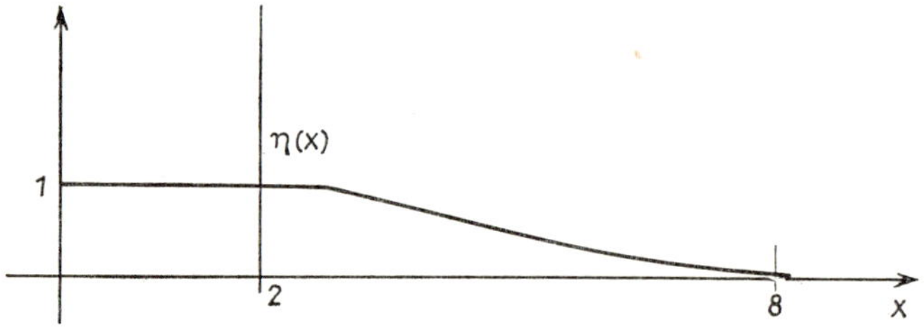

We also assume that

(8a) $$\eta' > -\tfrac{1}{2}.$$

Let us denote by p, ϕ, U, P, Q, u, v a particular (but not specified) set $p_i, \phi_i, U_i, P_i, Q_i, u_i$, $i = 1, \ldots, n$. Let us agree, moreover, that whenever x and ϕ, U, P, Q, u or v appear in the same formula, then the choice is determined by $x \in U$, and u, v mean $u(x)$, $v(x)$, respectively.

Define Λ by

(8b) $$\begin{cases} \Lambda(x) = \tfrac{3}{2}\,\lambda(u^2)\,\eta(v^2) & \text{if } x \in U_1 \cup \cdots \cup U_n; \\ \Lambda(x) = 0 & \text{otherwise.} \end{cases}$$

Clearly, Λ is smooth (because $\Lambda = 0$ outside ϕ^{-1} (ball of radius 3)) and non-negative.

Put $g = f - \Lambda$.

9. *Proof of the desired properties* (6a) – (6k)

Properties 6a, b and c are clearly satisfied. Moreover, since $f = u^2 - v^2$, if $f \geqq 2$, necessarily $u^2 \geqq 2$ and then $\lambda(u^2) = 0$. Thus we have (6d).

Since $\nabla g = \nabla f - \nabla \Lambda$, $x \in M$ is a critical point of g if and only if $(\nabla f)_x = (\nabla \Lambda)_x$.

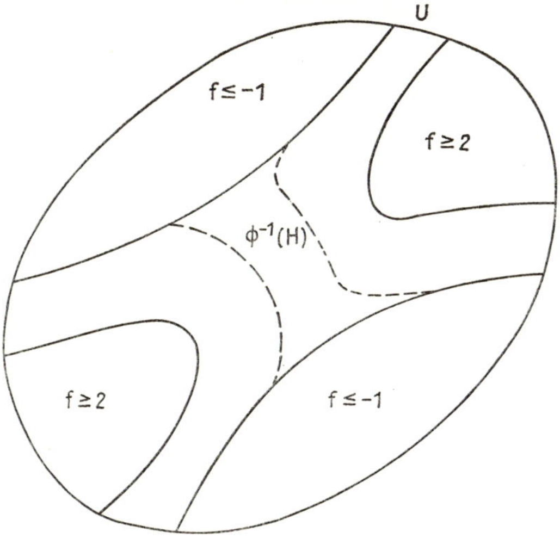

Now, for $x \in [-1 \leqq f \leqq +2]$ but outside $U_1 \cup \cdots \cup U_n$, we have $(\nabla \Lambda)_w = 0$ and $(\nabla f)_x \neq 0$, because the only critical points of f are $p_1, \ldots, p_n$.

That means that g (like f) has no critical points in $[-1 \leqq f \leqq 2] - (U_1 \cup U_2 \cdots \cup U_n)$.

Assume now that $x \in U$, $x \neq p$. Then the differential of g is

(9.1) $\quad g_*(x) = 2u\,(1 - \tfrac{3}{2}\lambda'(u^2)\,\eta(v^2))\,u_*(x) + 2v\,(-1 - \tfrac{3}{2}\lambda(u^2)\,\eta'(v^2))\,v_*(x).$

We shall verify that $g_*(x) \neq 0$.

In order to prove this we need the following statement.

(9.2) If $u(x)$ and $v(x)$ are both non-zero, then $u_*(x)$ and $v_*(x)$ are linearly independent.

In fact, from $u(x) \neq 0$, $v(x) \neq 0$ it follows that $e = P\phi x$ and $e_1 = Q\phi x$ are both different from $0 \in E$. Now consider the curves

$$\alpha(t) = \phi^{-1}(\phi(x) + te)$$
$$\beta(t) = \phi^{-1}(\phi(x) + te_1).$$

Then

$$u_*(x)\,\alpha'(0) = \frac{d}{dt}(u \circ \alpha(t))\bigg|_{t=0} = \frac{d}{dt}(\|P\phi\,(\phi^{-1}\,(\phi(x) + te))\|)\bigg|_{t=0}$$
$$= \frac{d}{dt}\,\|(1 + t)\,e\| = \|e\| \neq 0.$$

Similarly we may obtain

$$u_*(x)\,\beta'(0) = 0$$

$$v_*(x)\,\alpha'(0) = 0$$

$$v_*(x)\,\beta'(0) = 2\|e_1\|^2 \neq 0.$$

That proves that $u_*(x)$ and $v_*(x)$ are linearly independent, establishing (9.2).

Assume now that $g_*(x) = 0$. If $u = 0$, then by (9.1),

$$g_*(x) = 2v\,(-1 - \tfrac{3}{2}\,\lambda(u^2)\,\eta'(v^2))\,v_*(x) = 0.$$

Since $x \neq 0$, and $u = 0$, it must be that $v_*(x) \neq 0$, and from $\eta'(v^2) > -\frac{1}{2}$, $0 \leqq \lambda \leqq 1$, it follows that $2v(-1-\frac{3}{2}\,\lambda(u^2)\,\eta'(v^2)) = 0$ is only satisfied when $v = 0$ also, contradicting $x \neq p$.

The case $v = 0$ is treated similarly.

Finally, if $u \neq 0$, $v \neq 0$, then $u_*(x)$ and $v_*(x)$ are linearly independent by (9.2) and $g_*(x) = 0$ implies $2v(-1-\frac{3}{2}\lambda(u^2)\,\eta'(v^2)) = 0$, which we have seen to be true only if $v = 0$, while we are assuming $v \neq 0$. Thus g has no critical points other than $p_1, \ldots, p_n$, which are, indeed, critical because Λ is constant on $u < \sqrt{2}/2$, $v < \sqrt{2}$, whence $(\nabla g)_p = (\nabla f)_p = 0$.

Thus, property (6h) is satisfied by Λ.

In order to establish (6k) we observe that outside $U_1 \cup \cdots \cup U_n$, f and g coincide by (6b). We shall consider each U_i separately and prove that

$$U \cap [f \leqq -1] \bigcup_h H = U \cap [g \leqq -1]. \tag{9.3}$$

Clearly this implies (6k).

Of course we shall deal with (9.3) after transposing it into E by means of ϕ.

Put $X = P(E)$, $Y = Q(E)$, $Q = 1 - P$, and $B =$ open ball of radius 10 around $0 \in E$. Clearly $E = X \oplus Y$, $X \perp Y$. We shall let x, y denote elements in X and Y respectively and x^2, y^2 denote $x^2 = |x|^2$, $y^2 = |y|^2$. If e, x and y appear in the same formula, it must be understood that $e \in E$, $e = x + y$ and $x \in X$, $y \in Y$. Applying ϕ, statement (9.3) may be written as follows.

(9.4) In B, the submanifolds:

$$V = [x^2 - y^2 \leqq -1] \tag{9.4.1}$$

$$W = [x^2 - y^2 - \tfrac{3}{2}\,\lambda(x^2)\,\eta(y^2) \leqq -1] \tag{9.4.2}$$

satisfy

(9.4.3) $$V \underset{h}{\cup} H = W,$$

where H is a (k, l) handle.

The proof depends on elementary (though sly) computations. First of all we observe that W may be defined more conveniently as

(9.5) $$W = [x^2 - y^2 - \tfrac{3}{2}\lambda(x^2) \leqq -1].$$

In fact, if $y^2 < 2$, $\eta(x^2) = 1$ and then

$$x^2 - y^2 - \tfrac{3}{2}\lambda(x^2)\,\eta(y^2) = x^2 - y^2 - \tfrac{3}{2}\lambda(x^2);$$

if $y^2 > 2$ and $x^2 > 1$, then

$$x^2 - y^2 - \tfrac{3}{2}\lambda(x^2)\,\eta(y^2) = x^2 - y^2 - \tfrac{3}{2}\lambda(x^2) = x^2 - y^2;$$

finally, if $y^2 > 2$ and $x^2 < 1$, then necessarily $x^2 - y^2 \leqq -1$ and consequently

$$x^2 - y^2 - \tfrac{3}{2}\lambda(x^2)\,\eta(y^2) \leqq -1,$$

$$x^2 - y^2 - \tfrac{3}{2}\lambda(x^2) \leqq -1.$$

This proves (9.5).

Define $K \subset B$ by

(9.6) $$K = \{e \in B;\ x^2 + 1 - \tfrac{3}{2}\lambda(x^2) \leqq y^2 \leqq x^2 + 1\},$$

so that K is the set of elements $e \in B$ satisfying

(9.6.1) $$x^2 + 1 - \tfrac{3}{2}\lambda(x^2) \leqq y^2,$$

and

(9.6.2) $$y^2 \leqq x^2 + 1.$$

Let D^k and D^l be the unit balls of Y and X respectively and define $h\colon D^k \times D^l \to K$ by

(9.7) $$h(y, x) = (\sigma(y^2))^{1/2}\, x + (1 + \sigma(y^2)\, x^2)^{1/2}\, y,$$

where σ is the smooth function defined as follows: if $0 \leqq t \leqq 1$, $\sigma(t)$ is the unique solution in $[0, 1]$ of

(9.8) $$\frac{3}{2}\,\frac{\lambda(\sigma(t))}{1 + \sigma(t)} = 1 - t.$$

This mapping is smooth and has the form shown in the following graph.

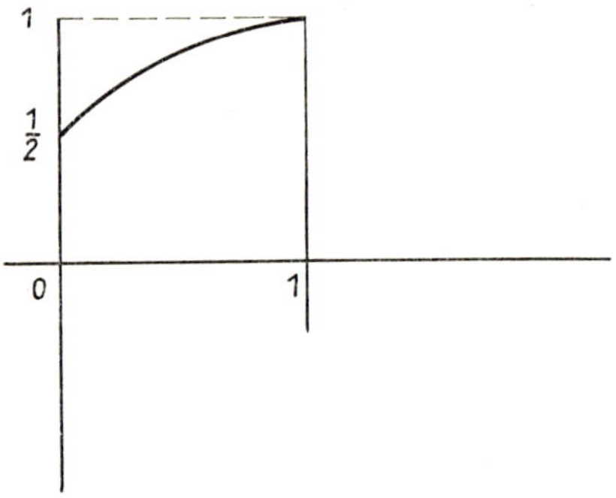

Since σ is smooth, h is also smooth. The reader will check that the image

$$H = h\,(D^k \times D^l)$$

of h is *contained in* K; this follows from the inequality

$$\sigma(y^2) + 1 - \tfrac{3}{2}\lambda\,(\sigma(y^2)\,x^2) \leqq (1 + \sigma(y^2)\,x^2)\,y^2$$

which is a consequence of

$$1 - y^2 = \tfrac{3}{2}\lambda\,(\sigma(y^2))\,(1 + \sigma(y^2))^{-1} \leqq \tfrac{3}{2}\,(\lambda(\sigma(y^2)\,x^2)\,(1 + \sigma(y^2)\,x^2)^{-1}$$

it follows from this inequality that $H \subset W$; the other condition ($y^2 \leqq 1 + x^2$) is even easier and is left to the reader.

Now consider the function S: $H \to X \times Y$ defined by:

$$(9.9) \qquad S(e) = \left((1 + x^2)^{-1/2}\,y,\ \left[\sigma\left(\frac{y^2}{1 + x^2}\right)\right]^{-1/2} x\right);$$

S is smooth and clearly Sh = identity, hS = identity.

In order to finish the proof it suffices to show that H is a (k, l) handle and that

$$V \underset{h}{\cup} H = W.$$

To do this, we first show that

$$H = K \cap [x^2 \leqq 1].$$

In fact, it is obvious that $H \subset K \cap [x^2 \leqq 1]$.

Assume now that $e = x + y \in K$ and $x^2 \leqq 1$. If $x \leqq \frac{1}{2}$, then $x^2 \leqq \sigma\,(y^2/(1 + x^2))$ is trivial. If $x^2 \geqq \frac{1}{2}$, then there exists $0 \leqq t \leqq 1$ such that $\sigma(t) = x^2$. From the formula

$$x^2 + 1 - \tfrac{3}{2}\lambda(x^2) = x^2 + 1 - \tfrac{3}{2}\lambda\,(\sigma(t)) \leqq y^2$$

it follows (use (9.8)) that:

$$x^2 + 1 - (1 + \sigma(t))(1 - t) \leqq y^2,$$

i.e.

$$(1 + x^2) - (1 + x^2)(1 - t) \leqq y^2,$$

so that

$$(1 + x^2)\, t \leqq y^2,$$

whence $t \leqq y^2/(1 + x^2)$. But since σ is increasing, we have

$$x^2 = \sigma(t) \leqq \sigma\left(\frac{y^2}{1 + x^2}\right),$$

and this shows that

(i) $$\left|\left[\sigma\left(\frac{y^2}{1 + x^2}\right)\right]^{-1/2} x\right| \leqq 1.$$

On the other hand, it is obvious from $y^2 \leqq x^2 + 1$ that

(ii) $$|(1 + x^2)^{-1/2}\, y| \leqq 1.$$

Now (i) and (ii) together imply that $S(e) \in D^k \times D^l$ and hence

$$e = hS(e) \in h(D^k \times D^l) = H.$$

This proves that $H \supset K \cap [x^2 \leqq 1]$ and we conclude that $H = K \cap [x^2 \leqq 1]$.

As a corollary, it follows that

(9.10) $$H \text{ is closed}$$

(This is the required Property 4.84, 2 from the definition of "handle".)

We now show that

(9.11) $$V \cup H = W.$$

(This is Property 4.84, 2a).

Plainly $V \cup H \subset W$. Let $e = x + y$ be a point such that

(a) $$x^2 - y^2 - \tfrac{3}{2}\lambda(x^2) \leqq -1.$$

If e is not in V we have

(b) $$x^2 - y^2 > -1.$$

(a) and (b) together imply that $\lambda(x^2) > 0$, whence

(c) $$x^2 < 1.$$

Now (a), (b) and (c) imply that e belongs to $K \cap [x^2 \leqq 1] = H$, and we obtain the desired assertion $W \subset V \cup H$.

If $e \in H \cap V \leqq K \cap V$, plainly $x^2 - y^2 = -1$, so $H \cap V \leqq \partial V$ and $y^2(1+x^2)^{-1} = 1$ verifying (4.84.2c).

Similarly, if $e \in H - M$ then $x^2 - y^2 > -1$, so that $y^2(x^2+1)^{-1} < 1$, which with (9.9) and (9.7) verifies (4.84.2d–2e). This completes the proof of (9.3) and of theorem (4.87). Q.E.D.

E. The Morse Inequalities

We begin with a rapid review of homology theory. Our description will be based on the axiomatic characterization of the homology groups, as given for example in Eilenberg-Steenrod ([4]).

We denote by (X, Y) a pair of topological spaces, Y a subspace of X; $(X, \varnothing)$ is written as X. Under suitable restrictions on the class of spaces considered, we can associate with each pair (X, Y) Abelian groups $H_k(X, Y)$, k an integer, ($H_k = \{0\}$ if $k < 0$). These groups will depend on a fixed group G (the "coefficient group" of the theory), so we should denote them by $H_k(X, Y; G)$. We are mainly interested in two specific cases, namely G = integers = Z or G = real numbers = R. In the case $G = R$ the H_k are vector spaces (over R) and the *Betti numbers* $\beta_k(X, Y)$ of the pair X, Y are defined by

$$\beta_k(X, Y) = \dim H_k(X, Y; R) \tag{1.1}$$

We write

$$\phi : (X, Y) \to (\bar{X}, \bar{Y})$$

if ϕ is continuous, $\phi : X \to \bar{X}, \phi(Y) \subseteq \bar{Y}$.

Any such function induces a homomorphism for each k

$$\phi^* : H_k(X, Y) \to H_k(\bar{X}, \bar{Y})$$

having the following properties:

(a) $(\phi\psi)^* = \phi^*\psi^*$

(b) if i = identity, i^* = identity.

(We can already deduce that homeomorphic pairs have isomorphic groups.)

(c) Let $\phi, \psi : (X, Y) \to (\bar{X}, \bar{Y})$ be homotopic

(here $\phi_t\ (X, Y) \to (\bar{X}, \bar{Y})$). Then $\phi^* = \psi^*$.

We say that (X, Y) and $(\bar{X}, \bar{Y})$ are *homotopically equivalent* iff there exist $\phi : (X, Y) \to (\bar{X}, \bar{Y})$ and $\psi : (\bar{X}, \bar{Y}) \to (X, Y)$ s.t. $\psi\phi$ and $\phi\psi$ are homotopic to the respective identity mappings. Property (c) implies that homotopically equivalent pairs have isomorphic homology groups.

(d) *Excision property*: Take a pair (X, Y), let $\mathcal{O} \subseteq Y$ be an open set such that also $\bar{\mathcal{O}} \subseteq \mathring{Y}$. Let $i =$ identity map from $(X - \mathcal{O}, Y - \mathcal{O})$ to (X, Y). Then

$$i^*: H_k(X - \mathcal{O}, Y - \mathcal{O}) \to H_k(X, Y)$$

is an *isomorphism onto*.

(e) The homology groups are related to the coefficient group G by

$$H_k(P) = \begin{cases} G & \text{if } k = 0 \\ \{0\} & \text{if } k > 0, \end{cases}$$

where $P = (P, \phi)$ is a space consisting of a single point.

(f) There exist maps $\partial_k: H_k(X, Y) \to H_{k-1}(Y)$,
(which we write simply as ∂, omitting the subindex) such that, if ϕ: $(X, Y) \to (\bar{X}, \bar{Y})$
Then

$$\partial \phi^* = (\phi | Y)^* \partial$$

(here we have designated $\phi | Y$ the restriction of ϕ to Y, $(\phi | Y)^*$ the induced map on $H_k(Y) = H_k(Y, \phi)$).

(g) *Exactness principle of Euler*

Let $X \supseteq Y \supseteq \varnothing$; let

$$j: (X, \varnothing) \to (X, Y)$$
$$k: (Y, \varnothing) \to (X, \varnothing)$$

be inclusion maps, j^*, k^* the induced maps on the homology groups. We can construct the sequence

$$\longrightarrow H_k(X) \xrightarrow{j_*} H_k(X, Y) \xrightarrow{\partial} H_{k-1}(Y) \xrightarrow{k_*} H_{k-1}(X) \xrightarrow{j_*} H_{k-1}(X, Y) \longrightarrow$$
$$\cdots \longrightarrow H_0(Y) \xrightarrow{k_*} H_0(X) \xrightarrow{j_*} H_0(X, Y) \xrightarrow{\partial} 0 \longrightarrow 0 \longrightarrow \cdots$$

(the *homology sequence* of the pair X, Y). The *exactness principle* tells us that the homology sequence of any pair (X, Y) is *exact* (i.e. the image of any group in the sequence under the corresponding homomorphism is equal to the kernel of the next homomorphism).

We note the following results for future use.

1. Let S^n be the n-sphere, $G = Z$, $n \neq 0$. Then

$$H_k(S^n) = 0 \quad \text{if} \quad k > 0, \quad k \neq n,$$
$$H_n(S^n) = Z$$
$$H_0(S^n) = Z$$

2. Let G be as before, D be the n-disk. Then

$$H_k(D^n, S^{n-1}) = H_k(D^n, \partial D^n) = \{0\} \quad \text{if} \quad k \neq n,$$

$$H_n(D^n, S^{n-1}) = Z.$$

3. Suppose $(X, Y) = \bigcup_{i=1}^{l} (X_i, Y_i)$, all X_i disjoint.

Then $H_k(X, Y) = \sum_{i=1}^{l} \oplus H_k(X_i, Y_i)$.

Example (i). To illustrate the use of the exactness principle we will deduce 2) from 1) and (e). Consider

$$H_k(S^{n-1}) \xrightarrow{k_*} H_k(D^n) \xrightarrow{j_*} H_k(D^n, S^{n-1}) \xrightarrow{\partial} H_{k-1}(S^{n-1}) \xrightarrow{k_*} H_{k-1}(D^n).$$

Since D^n is homotopic to a point for any n, we have $H_k(D^n) = \{0\}$ for any $k \neq 0$. Therefore, if $k > 0$, $k \neq n$, we get the sequence

$$\{0\} \longrightarrow H_k(D^n, S^{n-1}) \xrightarrow{\partial} \{0\} \longrightarrow \{0\},$$

whose exactness implies readily that $H_k(D^n, S^{n-1}) = \{0\}$.

On the other hand, if $k = n$, the sequence is

$$\{0\} \to H_n(D^n, S^{n-1}) \to Z \to \{0\}.$$

This time, exactness implies that $H_n(D^n, S^{n-1}) = Z$.

Example (ii). We now note a result more general than the exactness principle. Let $X \supseteq Y \supseteq Z$; consider the inclusion maps

$$j: (X, Z) \to (X, Y)$$

$$k: (Y, Z) \to (X, Z)$$

$$l: (Y, \varnothing) \to (Y, Z).$$

Using the induced mappings j^*, k^*, l^* we can form the sequence:

$$\longrightarrow H_k(X, Z) \xrightarrow{j^*} H_k(X, Y) \xrightarrow{\partial'} H_{k-1}(Y, Z) \xrightarrow{k^*} H_{k-1}(X, Z)$$

$$\xrightarrow{j^*} H_{k-1}(X, Y) \xrightarrow{\partial'} \cdots \xrightarrow{\partial'} H_0(Y, Z) \xrightarrow{k^*} H_0(X, Z) \xrightarrow{j^*} H_0(X, Y)$$

$$\xrightarrow{\partial'} \{0\} \xrightarrow{k^*} \{0\} \xrightarrow{j^*} \cdots \tag{1.2}$$

where ∂' is the composition of l^* and ∂, i.e.

$$\begin{array}{ccc} H_{k-1}(Y) & \xrightarrow{l^*} & H_{k-1}(Y, Z) \\ \uparrow \partial & \nearrow \partial' = l^* \partial & \\ H_k(X, Y) & & \end{array} \tag{1.3}$$

As an exercise the reader should prove, using the exactness principle, that (1.2) is an exact sequence.

Consider now the case $G = R$; the homology groups are then vector spaces. Let $K_k(X, Y)$ be the subspace of $H_k(X, Y)$ which is either the image of the preceding homomorphism or the kernel of the next, (similarly define $K_k(X, Z), \ldots$ etc.) and set $\varepsilon_k(X, Y) = \dim K_k(X, Y) \ldots$ etc. From the exactness of (1.2) we easily see that

$$\begin{cases} \beta_k(X, Y) = \varepsilon_k(X, Y) + \varepsilon_{k-1}(Y, Z) & \text{(1.4a)} \\ \beta_k(Y, Z) = \varepsilon_k(Y, Z) + \varepsilon_k(X, Z) & \text{(1.4b)} \\ \beta_k(X, Z) = \varepsilon_k(X, Z) + \varepsilon_k(X, Y) & \text{(1.4c)} \end{cases}$$

Now,

$$\begin{aligned} &\sum_{j \leqq m} (-1)^j \beta_j(X, Z) - \sum_{j \leqq m} (-1)^j \beta_j(X, Y) - \sum_{j \leqq m} (-1)^j \beta_j(X, Z) \\ &= \sum_{j \leqq m} (-1)^j \{\varepsilon_j(X, Z) + \varepsilon_j(X, Y) - \varepsilon_j(X, Y) - \varepsilon_{j-1}(Y, Z) \\ &\quad - \varepsilon_j(Y, Z) - \varepsilon_j(X, Z)\} \qquad \text{(1.5)} \\ &= -\sum_{j \leqq m} (-1)^j (\varepsilon_j(Y, Z) + \varepsilon_{j-1}(Y, Z)) = (-1)^{m+1} \varepsilon_m(Y, Z). \end{aligned}$$

Now define

$$\eta_m(X, Z) = (-1)^m \sum_{j \leqq m} (-1)^j \beta_j(X, Z). \tag{1.6}$$

Clearly it follows that

$$\begin{aligned} &\eta_m(X, Z) = \eta_m(X, Y) + \eta_m(Y, Z) - \text{non-negative integer, i.e.} \\ &\eta_m(X, Z) \leqq \eta_m(X, Y) + \eta_m(Y, Z). \end{aligned} \tag{1.8}$$

We now apply these results to Morse theory. Let M be a Hilbert manifold, f a smooth function on M satisfying the P-S condition on $a \leqq f \leqq b$; let c, $a < c < b$, be its only critical level, and suppose that the critical points of f $p_1, \ldots, p_n$ are non-degenerate, their indices being (n_i, m_i), $i = 1, \ldots, n$.

We know, by Theorem 4.87, that

$$[f \leqq b] \sim [f \leqq a] \underset{h_1}{\cup} H_1 \cup \cdots \underset{h_n}{\cup} H_n \tag{1.9}$$

Our aim is to compute $H_k([f \leqq b], [f \leqq a])$.

Theorem 4.88.

$$H_k(\{f \leqq b\}, \{f \leqq a\}) = \sum_{i=1}^{n} \oplus \delta_{kn_i} Z \tag{1.10}$$

Before beginning the proof, let us note the surprising fact (R. S. Palais ((3), p. 336)) that according to Theorem 4.88 the homotopy type of the pair $(\{f \leqq b\}, \{f \leqq a\})$ will depend only on the critical points with *finite* index at intermediate levels, those with infinite index being "homotopically invisible". This unexpected fact saves us from making the rather inelegant assumption of finiteness of the indices.

On to the proof! By Theorem 4.87, we have

$$\{f \leqq b\} \cong \{f \leqq a\} \underset{h_1}{\cup} H_1 \underset{h_2}{\cup} H_2 \cup \cdots \underset{h_n}{\cup} H_n,$$

$h_i: D^{n_i} \times D^{m_i} \to H_i$. Let R^{n_i} denote D^{n_i} with a ball removed from its interior, and let $\mathscr{R}^{n_i} = h_i(R^{n_i} \times D^{m_i})$. It is easy to see that the pair $(\{f \leqq b\}, \{f \leqq a\})$ is homotopic to $(\{f \leqq b\}, \{f \leqq a\} \cup \mathscr{R}^1 \cup \cdots \cup \mathscr{R}^n)$. By excision, this pair will have the same homology groups as $(\{f \leqq b\} - \{f < a\}, \mathscr{R}^1 \cup \cdots \cup \mathscr{R}^n)$ and also the same groups as $(H_1 \cup \cdots \cup H_n, \mathscr{R}^1 \cup \cdots \cup \mathscr{R}^n)$. But by a previous observation, the homology groups of the last pair are

$$\Sigma H_k(H_i, \mathscr{R}_j) = \Sigma H_k(D^{n_i}, \partial D^{n_i}).$$

By another of our observations,

$$H_k(D^{n_i}, \partial D^{n_i}) = \delta_{kn_i} Z, \quad n_i < \infty,$$

and the same formula holds for $n_i = \infty$, since D^{n_i} is homotopically trivial modulo ∂D^{n_i} in this case. This completes the proof.

Corollary: $\beta_k(\{f \leqq b\}, \{f \leqq a\})$ = number of critical points of type (k, ∞) between a and b.

We can now obtain the best known results in Morse theory.

Theorem 4.89 (Morse Inequalities): *Let M be a complete Hilbert manifold, f a smooth function satisfying the P-S condition on $a \leqq f \leqq b$, and suppose that a, b are regular values of f and that all critical points of f are non-degenerate. For each non-negative integer m let β_m be the m-th Betti number of the* *pair* $(\{f \leqq b\}, \{f \leqq a\})$, *and let* c_m *denote the number of critical points of* f *of index* (m, ∞) *in* $f^{-1}([a, b])$. *Then*

$$\begin{aligned} \beta_0 &\leqq c_0 \\ \beta_1 - \beta_0 &\leqq c_1 - c_0 \\ &\vdots \\ \sum_{m=0}^{k} (-1)^{k-m} \beta_m &\leqq \sum_{m=0}^{k} (-1)^{k-m} c_m \end{aligned} \tag{1.11}$$

and

$$\sum_{m=0}^{\infty} (-1)^m \beta_m = \sum_{m=0}^{\infty} (-1)^m c_m. \tag{1.12}$$

Corollary 1. $\beta_m \leqq c_m$ *for all* m.

Corollary 2. *If* f *is bounded below, then the conclusion of the theorem and of Corollary* 1 *remain valid if we interpret* β_m = *m-th Betti number of* $[f \leqq b]$ *and* c_m = *number of critical points of* f *having index* m *in* $\{f \leqq b\}$ *respectively.*

Proof of Theorem 4.89. Let $c_1 < c_2 < \cdots < c_n$ be the critical values of f in $[a, b]$. Choose a_i, $i = 0, 1, \ldots, n$, so that $a = a_0 < c_1 < a_1 < c_2 < \cdots < a_{n-1} < c_n < a_n = b$ and call $X_i = \{f \leqq a_i\}$. Then by the corollary to Theorem 4.88 it follows that $\beta_k(X_{i+1}, X_i)$ = number of critical points of index k in the level c_i. We thus have (see (1.6))

$$\begin{aligned} \eta_m(X_{i+1}, X_i) &= \sum_{k \leqq m} (-1)^{k-m} \beta_k(X_{i+1}, X_i) \\ &= \sum_{k \leqq m} (-1)^{k-m} \quad \text{(number of critical points of index } k \text{ on level } c_i). \\ \sum_{k \leqq m} \eta_k(X_n, X_0) &= \sum_{k \leqq m} (-1)^{k-m} \quad \text{(number of critical points of index } k \text{ in } f^{-1}([a, b])). \end{aligned}$$

But an iteration of (1.8) yields

$$\eta_m(\{f \leqq b\}, \{f \leqq a\}) \leqq \sum_{k \leqq m} \eta_k(X_{k+1}, X_k)$$

and by definition of η_m, (1.11) follows. Equation (1.12) can be obtained by taking m large enough.

The proof of Corollary 1 is trivial, and the proof of 2 follows easily from Theorem 4.89.

Bibliography

1. Lang, Serge, *Introduction to Differential Manifolds* (Interscience, John Wiley, New York, 1962).
2. Palais, Richard, "Morse theory on Hilbert manifolds", Topology Vol. **2**, pp. 299–340 (Dec. 1963).
3. Palais, R. and Smale, S., "A generalized Morse theory", Bull. A.M.S. Vol. **70**, pp. 165–172 (January 1964).
4. Eilenberg, Steenrod, *Foundations of Algebraic Topology* (Princeton University Press, 1953).
5. Milnor, *Morse Theory* (Princeton University Press, 1963).

CHAPTER V

Category

A. Definition and Elementary Properties

Remark: We introduce our concepts for arbitrary topological spaces X, but in the applications X will be a manifold.

5.1. Definition: *The closed set $A \subseteq X$ is said to be of first category with respect to X (in symbols $cat_X(A) = 1$ or simply $cat(A) = 1$) iff the injection $i: A \to X$ is homotopic to a constant.*

5.2. Definition: *The closed set $A \subseteq X$ is said to be of kth category with respect to X ($cat_X(A) = cat(A) = k$) iff*

(a) $A = A_1 \cup A_2 \cup \cdots \cup A_k$,
A_i closed and $\text{cat}(A_i) = 1$, $1 \leqq i \leqq k$.

(b) If $A = A_1 \cup A_2 \cup \cdots \cup A_l$, A_i, $1 \leqq i \leqq l$ closed and $l < k$, then some A_i is *not* of the first category.

The closed set $A \subseteq X$ is said to have infinite category with respect to X ($\text{cat}_X(A) = \text{cat}(A) = \infty$) if no decomposition of the form (a) in Def. (5.2) is possible.

Examples: Let E^n, E^∞ denote the Euclidean n-space and separable Hilbert space respectively. Then, if $A = \{x \mid |x| = 1\}$,

$$\text{cat}_{E^n}(A) = 1, \quad 1 \leqq n \leqq \infty$$

$$\text{cat}_A(A) = \begin{cases} 2 & \text{for} \quad 1 \leqq n < \infty \\ 1 & \text{for} \quad n = \infty. \end{cases}$$

We now prove some elementary properties of the notion of category.

5.3. Lemma: Let A, B closed subsets of X. Then

(5.3.1) $\quad \text{cat}(A \cup B) \leqq \text{cat}(A) + \text{cat}(B)$

(5.3.2) $\quad A \subseteq B \quad \text{implies} \quad \text{cat}(A) \leqq \text{cat}(B)$

(5.3.3) Let I be the closed unit interval, $\eta: X \times I \to X$ a continous function such that $\eta(x, 0) = x$ for $x = X$. Then if $\eta(x, 1) = \eta_1(x)$, $\text{cat}(A) \leqq \text{cat}(\eta_1(A))$.

Proof: (5.3.1.) and (5.3.2) are trivial. To prove (5.3.3.) we can evidently suppose that cat $(\eta_1(A))$ is finite, say, k. Then $\eta_1(A) = B_1 \cup B_2 \cup \cdots \cup B_k$, B_k closed and cat $(B_i) = 1$. Let $A_i = \eta_1^{-1}(B_i)$. Then $A = A_1 \cup \cdots \cup A_k$. The identity mapping in each A_i is clearly homotopic to a constant, so cat $(A) \leqq k$. Q.E.D.

5.4. Definition: We recall that the *dimension* of a compact metric space X (in symbols, dim X) is equal to n iff

(a) Every open covering $\{U_j\}$ of X has a refinement $\{V_j\}$ such that the order of $\{V_j\}$ is not greater than n, i.e., such that no $n + 2$ of the V_j have non-void intersections.

(b) There exists a covering $\{U_j\}$ such that for every refinement $\{V_j\}$ of $\{U_i\}$ there exist $V_{j_1} \cdots V_{j_{n+1}}$ in $\{V_j\}$ with non-void intersections (thus $\{V_j\}$ has order n).

A fundamental relation between category and dimension is given by

5.5. Theorem: *Let M be a Hilbert manifold, $A \subseteq M$ a compact set. Then*

(5.5) $$\text{cat}(A) \leqq \dim(A) + 1.$$

First we prove

5.6. Lemma: *Let M be a Hilbert manifold, $A \subseteq M$ a compact set: let $\eta: I \times A \to M$ be a homotopy between the identity mapping and a constant mapping* ($\eta(0, x) = x$, $\eta(1, x) =$ constant). *Then there is an extension $\bar{\eta}$ of η to $I \times \bar{U}$, U a neighborhood of A, such that η is again a homotopy between the identity mapping (in $\bar{U}$) and a constant.*

From Lemma 5.6 we deduce immediately:

Corollary: For A, M as in the lemma, suppose $\text{cat}_M(A) \leqq k$. Then there exists a neighborhood U of A such that $\text{cat}_M(\bar{U}) \leqq k$.

Proof of Lemma 5.6: Embed M as a regular submanifold of some Hilbert space H. Let $\Gamma \subset H \times H$ denote the set of all pairs (p, x), $p \in M$, $x \perp M_p$

(M_p the tangent space to M at p). Let $\pi : \Gamma \to H$, $\psi : \Gamma \to H$ be defined by

$$\pi(p, x) = p$$

$$\psi(p, x) = p + x$$

Let $[A] = \bigcup_{0 \leqq t \leqq 1} \eta(t, A) \supseteq A$. It is easy to see, using the implicit function theorem, that there exists a neighborhood V of $[A]$ and an $\varepsilon > 0$ such that, for $(p, x) \in \Gamma$, $p \in V$, $|x| < \varepsilon$, $\psi(p, x)$ covers a neighborhood $\mathcal{O}$ of $\bar{A}$ in H and is smoothly invertible there. Let ψ^{-1} be the inverse, and let $\phi = \pi\psi^{-1}$. Then ϕ is a smooth map from $\mathcal{O}$ into N, and $\phi|[A] =$ identity.

It is easy to see that η can be extended to $\hat{\eta}$ defined in $I \times M$ with values in H in such a way that $\hat{\eta}(0, x) = x$, $x \in M$, $\hat{\eta}(1, x) =$ constant. Choose a neighborhood U of A such that $\hat{\eta}(t, U) \subseteq \mathcal{O}$ and set

$$\bar{\eta}(t, p) = \pi\psi^{-1}\hat{\eta}(t, p), \quad p \in U, \quad 0 \leqq t \leqq 1.$$

Then $\bar{\eta}(0, p) = p$, $\bar{\eta}(1, p) =$ constant, $\bar{\eta}$ is M-valued, continuous and defined in $I \times U$.

Q.E.D.

Proof of Theorem 5.5: Let $\{U_j\}$ be any covering of A. Then there exists a refinement $\{V_k\}$ of $\{U_j\}$ such that $\text{cat}_M(\bar{V}_k) = 1$ for all k (for instance, a refinement consisting of coordinate patches).

Since $\dim A = n$ and by our observation, we can find an open covering $\{U_i\}$ such that

(a) $\text{cat}\, \bar{U}_i = 1$

(b) $\bigcap_{j=1}^{n+2} \bar{U}_{i_j} = \phi$ for any $(n + 2)$-plc of sets in $\{U_i\}$

(c) There exist $U_{i_1} \cdots U_{i_{n+1}}$ such that

$$\bigcap_{j=1}^{n+1} U_{i_j} \neq \phi.$$

Let $U = \bigcup \left\{ \bigcap_{1}^{n+1} U_{i_j} \middle| U_{i_j} \in \{U_i\} \right\}$, where the intersection is taken over sets $\{i_j\}$ of distinct indices.

By (c), $U \neq \phi$. By (a) each of the intersections has category 1, for it is contained in some U_i. Hence $\bar{U}$ itself has category 1—it is the disjoint union of sets of category 1. Observe that

$$A = (A - U) \cup U.$$

$A - U \subseteq \bigcup \{\bar{U}_i - U\}$ and it is easy to see that $\{\bar{U}_i - U\}$ satisfies (b) with $n + 1$ replacing $n + 2$. Indeed; since every possible intersection of $n + 1$ different sets $\bar{U}_j$ has been subtracted from $\bar{U}_i$, the intersection of $n + 1$ different sets $\bar{U}_i - U$ with different indices must be empty. Proceeding by induction (the case $n = 1$ being trivial) $\text{cat}_M(A) \leqq n + \text{cat}\, U \leqq n + 1$.

Q.E.D.

B. Category and Homology

We now define the singular homology groups of a topological space. These amount to a concrete realisation of the homology groups considered axiomatically at the end of Chapter 4. We will define cubical homology groups rather than the usual simplicial ones.

Let $I = [-1, +1]$.

5.7. Definition: A *singular n cube* in the topological space X is a continuous mapping $\phi : I^n \to X$.

5.8. Definition: A singular n-cube is called *degenerate* if ϕ does not depend on all of its coordinates.

If $\phi(x_1, \ldots, x_k, \ldots, x_n) = \phi(x_1, \ldots, x_k^1, \ldots, x_n)$ we say that ϕ is *degenerate along its k-th coordinate.*

Let $Q_n(X)$ be the free Abelian group generated by all the singular n-cubes in X. Let $D_n(X)$ be the free Abelian group generated by all degenerate n-cubes in X. Then

5.9. Definition: $C_n(X) = n$-th singular cubic chain group is defined as

$$C_n(X) = Q_n(X)/D_n(X).$$

Given a singular n cube ϕ, we obtain from it two $n - 1$ cubes (*the k-th faces of* ϕ), ϕ_k^+, ϕ_k^- as follows.

Given $(x_1 \cdots x_{k-1}, x_{k+1} \cdots x_n) \in I^{n-1}$, define

$$(x_1, \ldots, x_{k-1}, x_{k+1}, \ldots, x_n) \to \phi(x_1, \ldots, x_{k-1}, 1, x_{k+1}, \ldots, x_n)$$

and

$$(x_1, \ldots, x_{k-1}, x_{k+1}, \ldots, x_n) \to \phi(x_1, \ldots, x_{k-1}, -1, x_{k+1}, \ldots, x_n).$$

We can then define a boundary operator as follows. If ϕ is an n-cube

$$\partial\phi = \sum_{k=1}^{n} (-1)^k (\phi_k^- - \phi_k^+)$$

and ∂ is extended linearly to arbitrary n-chains. Since (as the reader may easily verify) $\partial: C_n(X) \to C_{n-1}(X)$, we can define a boundary ∂ in C

(Definition 5.7 should be compared)

$$\partial : C_n(x) \to C_{n-1}(X).$$

5.10. Lemma: $\partial\partial = 0$.

Proof: It is evidently sufficient to consider only singular n-cubes $\phi: I^n \to X$. We have

$$\partial\phi = \sum_{k=1}^{n} (-1)^k (\phi_k^- - \phi_k^+)$$

$$\partial\partial\phi = \sum_{k=1}^{n} (-1)^k \left(\sum_{l=1}^{n-1} (-1)^l (\phi_{kl}^{--} - \phi_{kl}^{-+}) \right)$$

$$- \sum_{k=1}^{n} (-1)^k \left(\sum_{l=1}^{n-1} (-1)^l (\phi_{kl}^{+-} - \phi_{kl}^{++}) \right)$$

$$= \sum_{k=1}^{n} \sum_{l=1}^{n-1} (-1)^{k+l} (\phi_{kl}^{--} - \phi_{kl}^{-+} + \phi_{kl}^{+-} - \phi_{kl}^{++}).$$

Observe now that if $l < k$

$$\phi_{kl}^{*0} = \phi_{l,k+1}^{0*}$$

where $(*, 0)$ stands for any combination of the signs $+$, $-$; and if $l > k$

$$\phi_{k,l+1}^{*0} = \phi_{l,k}^{0*}.$$

This shows that $\partial\partial\phi$ will be an element of $D_n(X)$, and hence $\partial\partial\phi = 0$ in C_n. Q.E.D.

Having defined chain groups and boundary operators we can define homology groups in the usual way, i.e.

5.11. Definition: Let $Z_n(X) = \ker \partial \subset C_n(X)$

$$B_n(X) = \partial C_{n+1}(X) \subset C_n(X).$$

Then $H_n(X) = n$-th (cubical) singular homology group of $X = Z_n(X)/B_n(X)$.

To define *relative* cubical groups of a pair (X, Y) offers no new difficulties; we simply define the n-th singular chain group of the pair (X, Y) as

$$C_n(X, Y) = C_n(X)/(D_n(X) + C_n(Y)),$$

and the boundary operator ∂ as the relativization of the former ∂ and the n-th cubical singular homology group of (X, Y) as

$$H_n(X, Y) = \frac{Z_n(X, Y) = \ker \partial}{B_n(X, Y) = \partial C_{n+1}(X, Y)}.$$

The cohomology groups are constructed as usual starting from the cochain groups and the coboundary operators. The reader may benefit by consulting

Hocking-Young or better still Hilton-Wylie, and by proving some of the Eilenberg-Steenrod axioms for our cubical singular groups.

It can be seen (H. and Y., p. 306, H. and W., p. 362, B., p. 110) that a multiplicative structure can be introduced into the cohomology groups, by means of the so-called cup product. Since we are mainly interested in cohomology groups of finite-dimensional manifolds, we will define this product in terms of differential forms—via De Rham's theorem.

5.12. Definition: Let M be an n-manifold, $Y \subseteq M$. Let $C^k(X, Y)$ denote the vector space of all smooth exterior forms on M of degree k which vanish in a neighborhood of Y. The coboundary operator is here simply the differential d as ordinarily defined for exterior forms.

$$d: C^k(X, Y) \to C^{k+1}(X, Y)$$

As usual $Z^k(X, Y) = \{\omega \in C^k(X, Y) : d\omega = 0\}$, $B^k(X, Y) = dC^{k-1}(X, Y)$ and

5.13. Definition: $H^k(X, Y) = Z^k(X, Y)/B^k(X, Y)$.

De Rham's theorem essentially states that these cohomology groups defined by means of differential forms coincide with the cohomology group of M defined by the singular cubical groups introduced above.

Since differential forms can be multiplied, the definition of cup product is already in view. Let $Y_1, Y_2 \subseteq M$, $\omega_1 \in C^{k_1}(M, Y_1)$, $\omega_2 \in C^{k_2}(M, Y_2)$.

Then $\omega_1 \wedge \omega_2 \in C^{k_1+k_2}(M, Y_1 \cup Y_2)$. Suppose that ω_1, ω_2 are cocycles, i.e. that $d\omega_1 = d\omega_2 = 0$. Then

$$d(\omega_1 \wedge \omega_2) = d\omega_1 \wedge \omega_2 + (-1)^{\deg \omega_1} \omega_1 \wedge d\omega_2 = 0.$$

If ω_1 is a cocycle and ω_2 is a coboundary so that $\omega_2 = d\omega$, then

$$d(\omega_1 \wedge \omega) = d\omega_1 \wedge \omega + (-1)^{\deg \omega_1} \omega_1 \wedge \omega_2 = \pm \omega_1 \wedge \omega_2,$$

which shows that $\omega_1 \wedge \omega_2$ is a coboundary. This allows us to define a product (we will use the sign $\cup$) in the cohomology groups, operating as follows:

$$H^{k_1}(M, Y_1) \times H^{k_2}(M, Y_2) \xrightarrow{\cup} H^{k_1+k_2}(X, Y_1 \cup Y_2).$$

We are going to establish now some rather interesting relations between the cohomology structure of a pair (M, Y) and the category of Y.

5.14. Theorem: *Let* cat $(X) = n$. *Then any cup product of n elements of degree > 0 vanishes.*

This theorem can be reformulated as follows. Define cuplength (X) = greatest number of elements of non-zero degree with nonvanishing cup product. Then we have the following corollary.

Corollary:

(5.6) $$\text{cat}\,(X) \geqq \text{cup length}\,(X) + 1.$$

Proof: Since the case $n = 1$ is evident (it reduces simply to the fact that, if the identity mapping $i: X \to X$ is homotopic to a constant, then X has trivial homology) we can suppose $n > 1$. Let $Y \subseteq X$ be a set of the first category, i.e. such that $i: Y \to X$ is homotopic to a constant map $m: Y \to X$, $m(y) \equiv p \in X$. Consider the exact cohomology sequence

$$H^{k-1}(Y) \longrightarrow H^k(X, Y) \xrightarrow{j^*} H^k(X) \longrightarrow H^k(Y).$$

Our assumption on Y implies that $H^k(Y) = 0$, whence by exactness, j_* is onto. Suppose now that X is of category n, $X = Y_1 \cup \cdots \cup Y_n$, Y_i of the first category. Let $\gamma_1, \ldots, \gamma_n$ be n elements in the cohomology rings of degrees $k_1, \ldots, k_n$. Then $\gamma_i = j^*\hat{\gamma}_i$, $\hat{\gamma}_i \in H^k(X, Y_i)$, $i = 1, \ldots, n$. The mapping j^* commutes with the cup product, so $(\gamma_1 \cup \gamma_2 \cup \cdots) = j^*(\hat{\gamma}_1 \cup \hat{\gamma}_2 \cup \cdots)$. But $\hat{\gamma}_1 \cup \hat{\gamma}_2 \cdots \in H^{k_1 + \cdots + k_n}(X, Y_1 \cup \cdots \cup Y_n) = H^{k_1 + \cdots + k_n}(X, X) = 0$.
Q.E.D.

Let X, $\hat{X}$ be two manifolds, ω a form on X. Then we can use ω in an evident way to define a form $\tilde{\omega}$ in $X \times \hat{X}$, and the same is true for forms in $\hat{X}$. Let π, $\hat{\pi}$ be the mappings of forms and of cohomology groups defined in this way.

$$\pi: H^k(X) \to H^k(X \times \hat{X})$$

$$\hat{\pi}: H^k(\hat{X}) \to H^k(X \times \hat{X})$$

It may be seen that $\pi H^k(X)$, $\hat{\pi} H^k(\hat{X})$ together generate the cohomology ring of $X \times \hat{X}$. But this proves the following statement.

5.15. Lemma:

(5.7) $$\text{cuplength}\,(X \times \hat{X}) \geqq \text{cuplength}\,(X) + \text{cuplength}\,(\hat{X}).$$

Applying this lemma to the torus T^n, cat $(T^n) - 1 \geqq$ cuplength $(T^n) \geqq n$. Indeed, equality holds here. The proof is left as an exercise.

Even though we have defined the cup product only for finite-dimensional manifolds, we have mentioned the fact that it can be defined in more general situations. We mention the following result, discussed in more detail later.

5.16. Theorem: *Let X be a finite dimensional simply connected manifold, $\Omega(X)$ its loop space. Then cuplength $\Omega(X) = \infty$.*

Corollary: cat $\Omega(X) = \infty$.

Example: If $P(\mathscr{H})$ is the Hilbert projective space over $\mathscr{H}$, then cat $P(\mathscr{H}) = \infty$.

C. Category and Calculus of Variations in the Large

Let f be a smooth function on a Hilbert manifold satisfying the P-S condition. Define $\Gamma_k(M)$ as the set of all subsets of M of category $\geqq k$.

5.17. Definition:

$$c_m(f) = \inf_{A \in \Gamma_m(M)} \left\{ \sup_{p \in A} f(p) \right\}$$

where we put $c_m(f) = \infty$ if $\Gamma_m(M) = \phi$.

Let $m < m'$. Then $c_m(f)$ is an infimum over more elements than $c_{m'}(f)$, therefore

$$-\infty \leqq c_1(f) \leqq c_2(f) \leqq \cdots \leqq \infty. \tag{5.8}$$

5.18 Lemma: *Suppose (as always) that the pair (M, f) satisfies Condition P-S. Let $\alpha(t)$ be a C^∞ real-valued function defined for $t \geqq 0$, such that $\alpha(t) \equiv 1$ for $0 \leqq t \leqq 1$, such that $t^2\alpha(t)$ is monotone increasing for $t \geqq 0$, and such that $t^2\alpha(t) = 2$ for $t \geqq 2$. Let $V(p) = -\alpha(|\nabla f(p)|)\nabla f(p)$, so that V is a C^1 tangent vector field on M, and let $\eta_t(p)$ be the flow defined by V. Then $\eta_t(p)$ is defined for all $p \in M$ and all $-\infty < t < +\infty$.*

Proof: It is plain from the description of the function α that the vector field $V(p)$ is uniformly bounded; let K be its upper bound. Since $d\eta_t(p)/dt = V(\eta_t(p))$, it follows from the definition of distance on the manifold M that $\varrho(\eta_t(p), \eta_s(p)) \leqq K|s - t|$ for $t_-(p) < s, t < t_+(p)$. Thus, if $t_+(p) < \infty$, and $\{t_n\}$ approaches $t_+(p)$ from below, $\{\eta_t/(p)\}$ is a Cauchy sequence, contradicting Proposition 4.46 in virtue of the completeness of M. It follows that $t_+(p) = \infty$. We may prove in the same way that $t_-(p) = -\infty$, and our lemma follows.

5.19 Lemma: *Let $\eta_t(p)$ be as in the preceding lemma. Let c be a real number, and set*

$$K_c = \{p \in M \mid f(p) = c, \quad (\nabla f)(p) = 0\}.$$

Then K_c is compact. Moreover, if for each $\varepsilon > 0$ we set

$$(5.9) \qquad N_\varepsilon = \{p \in M \,|\, |f(p) - c| < \varepsilon \quad \text{and} \quad |(\nabla f)(\eta_t(p))| < \varepsilon \quad \text{for some } t \text{ such that } 0 \leqq t \leqq 1\},$$

then any neighborhood U of K contains one of the neighborhoods N_ε of K.

Proof: The assertion that K_c is compact follows immediately from Condition P-S; thus only our second assertion requires proof. Suppose that this second assertion is false. The there exist a neighborhood U of K_c not containing any of the sets N_ε, and hence there exists a sequence of points $p_n \in M$, and a sequence t_n of numbers such that $0 \leqq t_n \leqq 1$, such that $f(p_n) \to c$ and $\nabla f(\eta_{t_n}(p_n)) \to 0$ as $n \to \infty$. Passing to a subsequence, we may suppose without loss of generality that $t_n \to t^*$ as $n \to \infty$. Now

$$(5.10) \qquad \frac{d}{dt} f(\eta_t(p)) = V(\eta_t(p)) f = -\alpha(|\nabla f(\eta_t(p))|)((\nabla f(\eta_t(p))) f)$$

and thus, from the definition of the gradient, we have

$$(5.11) \qquad \frac{d}{dt} f(\eta_t(p)) = -\alpha(|\nabla f(\eta_t(p))|)\,|\nabla f(\eta_t(p))|^2.$$

It is clear from (5.11) and from the definition of α that $|df(\eta_t(p))/dt|$ is uniformly bounded for all $p \in M$ and real t; thus there exists a finite constant K such that $|f(\eta_t(p)) - f(p)| \leqq K|t|$. Since $f(p_n) \to c$, it follows from this last that $f(\eta_{t_n}(p_n))$ is uniformly bounded. Thus, by Condition P-S, $\{\eta_{t_n}(p_n)\}$ has a convergent subsequence, and we may suppose without loss of generality that $\{\eta_{t_n}(p_n)\}$ converges to a point $q \in M$. Since $(\nabla f)(\eta_{t_n}(p_n)) \to 0$, q is evidently a critical point of f. We have

$$(5.12) \qquad p_n = \eta_{-t_n}(\eta_{t_n}(p_n)) \to \eta_{-t^*}(q) = q$$

by Lemma 4.45 and by the fact that q is a critical point. Thus $q \in K_c$ is the limit of $\{p_n\}$, and since $p_n \notin U$ we have a contradiction which completes our proof.

5.20 Corollary: *Let $0 < \varepsilon < 1$. Let η_t and N_ε be as in the preceding Lemma. Then if $f(p) \leqq c + \varepsilon^2/2$ and $p \notin N_\varepsilon$, we have $f(\eta_1(p)) \leqq c - \varepsilon^2/2$.*

Proof: It is plain from (5.11) of the preceding proof that $f(\eta_t(p))$ decreases as t increases, and, in fact, that

$$(5.13) \qquad f(\eta_1(p)) - f(p) = -\int_0^1 \alpha(|\nabla f(\eta_t(p))|)\,|\nabla f(\eta_t(p))|^2\, dt.$$

If $f(p) \leqq c - \varepsilon$ we have nothing to prove; thus we may suppose without loss of generality that $|f(p) - \varepsilon| < \varepsilon$. Then $p \notin N_\varepsilon$ implies that $|\nabla f(\eta_t(p))| \geqq \varepsilon$ for $0 \leqq t \leqq 1$, so that, since $t^2\alpha(t)$ is monotone increasing (cf. the first paragraph of the preceding proof) we may conclude, using (b), that $f(\eta_1(p) - f(p)) \leqq -\varepsilon^2$. But then

$$f(\eta_1(p)) \leqq c + \frac{\varepsilon^2}{2} - \varepsilon^2 \leqq c - \frac{\varepsilon^2}{2},$$

and the present corollary is proved.

We are now in a position to prove the principal theorem of Lusternik and Schnirelman in a generalized form.

5.21 Theorem: *Let (M, f) satisfy Condition P-S, and let $\{c_m(f)\}$ be as in Definition 5.17. Suppose that $m \leqq n$, and that $-\infty < c = c_m(f) = c_n(f) = c_n(f) < +\infty$. Then the set K_c of critical points (cf. (1)) is of category $n - m + 1$ at least; moreover, even if $m = n$, the set K_c is non-empty.*

5.22 Corollary: *Under the hypotheses of the preceding theorem the set K_c is of dimension $n - m$ at least.*

Proof: The corollary follows immediately from the theorem and Theorem 5.5. To prove the theorem, first suppose that $n > m$ and that cat $(K_c) \leqq n - m$, and then use the Corollary of Lemma 5.6 to find a neighborhood U of K_c such that cat $(\bar{U}) \leqq n - m$. Using Lemma 5.19 and Lemma 5.3.2, we may suppose without loss of generality that U is one of the neighborhoods N_ε described by (2). By Definition 5.17, there exists a closed subset A of M such that cat $(A) \geqq n$ and such that $\sup \{f(p) \mid p \in A\} \leqq c + \varepsilon^2/2$. Put $A_0 = A - N_\varepsilon$. Then, by Lemma 5.3.1, cat $(A_0) \geqq m$. Thus, if η_t is as in Lemma 5.19 and Corollary 5.20, it follows from Lemma 5.3.3 that cat $(\eta_1(A_0)) \geqq m$. On the other hand, by Corollary 5.20, $f(\eta_1(p)) \leqq c - \varepsilon^2/2$ for $p \in A_0$. This contradicts the Definition 5.17 of c_m and thus completes the proof of Theorem 5.21 in case $n > m$. In case $n = m$ and K_c is void, we may let U be the null set, and arrive by the same argument at the same contradiction. Thus Theorem 5.21 follows in every case. Q.E.D.

Reference

L. Liusternik and L. Schnirelman, *Méthodes topologiques dans les problèmes variationnels* (Hermann & Cie, Editeurs, Paris, 1934).

CHAPTER VI

Applications of Morse Theory to Calculus of Variations in the Large

Bibliography

1. R.S.Palais, "Morse theory on Hilbert Manifolds", *Topology*, Vol. 2, pp. 299–340.
2. J.Milnor, *Morse theory* (Ann. of Math. Studies, Princeton, 1963).
3. I.M.Singer, *Notes on Differential Geometry* (Mimeographed, M.I.T., 1962).
4. S.S.Chern, *Differentiable manifolds* (Mimeographed, Chicago Univ., 1959).

We consider now the set of all suitably smooth paths in a finite-dimensional compact Riemannian manifold M. We shall see that a natural (infinite-dimensional) Riemannian structure can be introduced into this set, allowing us to apply our infinite-dimensional Morse theory. Extremals of a conveniently chosen f on this set will correspond to geodesics in M, so that our results will relate to the geodesics of M.

6.1. Definition: *Let R^n be n-dimensional Euclidean space. Define* $H_0(I, R^n) = L^2(I, R^n)$, i.e. *the space of all functions, σ, ϱ, ... such that* $\int_0^1 |\sigma(t)|^2\, dt < \infty$ *with the scalar product*

$$(\sigma, \varrho)_0 = \int_0^1 (\sigma(t), \varrho(t))\, dt.$$

6.2. Definition: *Let $H_1(I, R^n)$ be the set of all absolutely continuous maps $\sigma : I \to R^n$ such that $\sigma' \in H_0(I, R^n)$. $H_1(I, R^n)$ is a Hilbert space under the inner product $(\sigma, \varrho)_1 = (\sigma(0), \varrho(0)) + (\sigma', \varrho')_0$. In fact, if $(p, q) \in R^n \oplus H_0(I, R^n)$, the map* $(p, q) \to p + \int_0^t g(s)\, ds \in H_1(I, R^n)$ *is an isometry onto.*

6.3. Definition: *We define* $L: H_1(I, R^n) \to H_0(I, R^n)$ *by* $L\sigma = \sigma'$ *and we define* $H_1^*(I, R^n) = \{\sigma \in H_1(I, R^n) \mid \sigma(0) = \sigma(1) = 0\}$.

Then the following is immediate:

6.4. Theorem: *L is a bounded linear transformation of norm 1. $H_1^*(I, R^n)$ is a closed linear subspace of codimension $2n$ in $H_1(I, R^n)$ and L maps $H_1^*(I, R^n)$ isometrically onto the set of $g \in H_0(I, R^n)$ such that*

$$\int_0^1 g(t)\,dt = 0,$$

i.e. *into the orthogonal complement in $H_0(I, R^n)$ of the set of constant maps of I into R^n.*

6.5. Theorem: *If $\varrho \in H_1^*(I, R^n)$ and λ is absolutely continuous from I into R^n, then*

$$\int_0^1 (\lambda'(t), \varrho(t))\,dt = (\lambda, -L\varrho)_0$$

6.6. Definition: $C^0(I, R^n)$ = *set of all continuous maps of I into R^n. $C^0(I, R^n)$ is a Banach space with the usual norm $|\ |_\infty$. The inclusion of $C^0(1, R^n)$ into $H_0(1, R_n)$ is evidently bounded.*

6.7. Theorem: *Let $\sigma \in H_1(1, R^n)$. Then*

$$|\sigma(t) - \sigma(s)| \leqq \sqrt{|t - s|}\,|L_\sigma|_0 .$$

Proof: Apply Schwarz's inequality.

Corollary 1: *If $\sigma \in H_1(1, R^n)$ then $|\sigma|_\infty \leqq 2\,|\sigma|_1$.*

Corollary 2: *The inclusion maps $i: H_1(I, R^n) \to C^0(I, R^n)$ and $H_0(I, R^n)$ are completely continuous.*

Proof of Corollary 1 is trivial. For 2 apply the Arzela-Ascoli Theorem.

6.8. Lemma: *Let $\phi: R^n \to R^p$ be a smooth map, and let $\bar{\phi}: H_1(I, R^n) \to H_1(I, R^p)$ be defined by $\bar{\phi}(\sigma) = \phi \circ \sigma$. Then $\bar{\phi}$ is smooth. Moreover, if $1 \leqq m \leqq k$, then*

$$d^m\bar{\phi}_\sigma(\lambda_1, \ldots, \lambda_m)(t) = d^m\phi_{\sigma(t)}(\lambda_1(t) \ldots \lambda_m(t)) .$$

This follows from

6.9. Lemma: *Let F be a C^1-map of R^n into $L^s(R^n, R^p)$, the space of all s-linear maps from $\pi^s R^n$ to R^p. Then the map F of $H_1(I, R^n)$ into*

$L^s(H_1(I, R^n), H_1(I, R^p))$ defined by

$$\bar{F}(\sigma)(\lambda_1 \ldots, \lambda_s)(t) = F(\sigma(t))(\lambda_1(t), \ldots, \lambda_s(t))$$

is continuous. Moreover, if F is C^3 then $\bar{F}$ is C^1 and

$$d\bar{F} = \overline{dF}.$$

Proof: Observe that

$$\bar{F}(\sigma)(\lambda_1 \cdots \lambda_s)'(t) = \frac{d}{dt} F(\sigma(t))(\lambda_1(t) \cdots \lambda_s(t)) = dF_{\sigma(t)}(\sigma'(t))(\lambda_1(t), \ldots, \lambda_s(t))$$
$$+ \sum_{i=1}^{s} F(\sigma(t))(\lambda_1(t), \cdots, \lambda_i'(t), \cdots, \lambda_s(t))$$

which implies

$$|\bar{F}(\sigma)(\lambda_1 \cdots \lambda_s)'(t)| \leqq |dF_{\sigma(t)}| \, |\lambda_1(t)| \, | \cdots |\lambda_s(t)|$$
$$+ \sum_i |F(\sigma(t))| \, |\lambda_1(t)| \cdots |\lambda_i(t)| \cdots |\lambda_s(t)|.$$

Since $|\lambda_i|_\infty \leqq 2|\lambda_i|_1$, and putting $k = \sup |dF_{\sigma(t)}|$, we have

$$|dF_{\sigma(t)}(\sigma'(t))(\lambda_1 \cdots \lambda_s)|_0 \leqq k 2^s L(\sigma) |\lambda_1|_1 \cdots |\lambda_s|_1.$$

Since also

$$|\bar{F}(\sigma)(\lambda_1 \cdots \lambda_i'(t) \cdots \lambda_s)| \leqq 2^s \sup |F(\sigma(t))| \, |\lambda_1|_1 \cdots \lambda_i'(t) \cdots |\lambda_s|_1,$$

if we recall that $|\varrho|_0^2 = |\varrho(0)|^2 + |\varrho'|_0^2$ we see that

$$|F(\sigma)(\lambda_1 \cdots \lambda_s)|_1 \leqq k(\sigma) |\lambda_1|_1 \cdots |\lambda_s|_1'$$

where $k(\sigma)$ is a constant depending on σ. It follows that (since $\bar{F}(\sigma)$ is plainly multilinear) $\bar{F}(\sigma) \in L^s(H_1(I, R^n), H_1(I, R^p))$. If $\varrho \in H_1(I, R^n)$ then

$$|(\bar{F}(\sigma) - \bar{F}(\varrho))(\lambda_1 \cdots \lambda_s)|_\infty \leqq 2^s \sup |F(\sigma(t)) - F(\varrho(t))| \, |\lambda_1|_1 \cdots |\lambda_s|_1$$

and it is plain that

$$|((\bar{F}(\sigma) - \bar{F}(\varrho))(\lambda_1 \cdots \lambda_s))'|_0 \leqq 2^s M(\sigma, \varrho) |\lambda_1|_1 \cdots |\lambda_s|_1,$$

where

$$M(\sigma, \varrho) = \sup |dF_{\sigma(t)}| \, |\sigma' - \varrho'|_0 + \sup |dF_{\sigma(t)} - dF_{\varrho(t)}| \, |\varrho'|_0$$
$$+ s \sup |F(\sigma(t)) - F(\varrho(t))|.$$

Hence

$$|\bar{F}(\sigma) - \bar{F}(\varrho)| \leqq k(\sigma, \varrho),$$

where $|\ |$ is the norm in $L^s(H_1(I, R^n), H_1(I, R^p))$ and where the constant $k(\sigma, \varrho) \to 0$ if $\sup |F(\sigma(t)) - F(\varrho(t))|$, $\sup |dF_{\sigma(t)} - dF_{\varrho(t)}|$ and $|\sigma' - \varrho'|_0$ all approach zero. But if $\varrho \to \sigma$ in $H_1(I, R^n)$ then $|\lambda' - \varrho'|_0 \leqq |\sigma - \varrho|_1$ goes to zero and then $\varrho \to \sigma$ uniformly. Hence since F and dF are continuous $F(\sigma(t)) \to F(\varrho(t))$ uniformly and $dF_{\varrho(t)} \to dF_{\sigma(t)}$ uniformly, so $k(\sigma, \varrho) \to 0$. This shows that $\bar{F}$ is continuous, and it can be proved similarly that $\bar{F}$ is C^1 whenever F is C^3.

Q.E.D.

Having disposed of the preliminaries we proceed to the applications.

6.10. Definition: *Let V be a finite dimensional smooth manifold. Denote by $H_1(I, V)$ the set of all continuous mappings $\sigma : I \to V$ such that $\phi\sigma$ is absolutely continuous and $|(\phi \circ \sigma)'|$ locally square-integrable for each chart ϕ in V. Let $H_1(I, V)_\sigma = \{\lambda \in H_1(I, T(V)) \mid \lambda(t) \in V_{\sigma(t)}$ for all $t \in I\}$, where $T(V)$ is the tangent bundle to V, and $V_{\sigma(t)}$ is the tangent space at $\sigma(t)$. If $p, q \in V$ we define $\Omega(V; p, q)$ as $\{\sigma \in H_1(I, V) \mid \sigma(0) = p,\ \sigma(1) = q\}$ and if $\sigma \in \Omega(V, p, q)$ we define $\Omega(V; p, q)_\sigma = \{\lambda \in H_1(1, V)_\sigma | \lambda(0) = 0_p,\ \lambda(1) = 0_q\}$ where 0_p (resp. 0_q) is the zero of V_p (resp. V_q).*

Remark: $H_1(I, V)_\sigma$ is a vector space under pointwise operations and $\Omega(V, p, q)_\sigma$ is a subspace of $H_1(I, V)_\sigma$.

6.11. Theorem: *Let V be a smooth submanifold of R^n. Then* (a) *$H_1(I, V)$ consists of all $\sigma \in H_1(I, R^n)$ such that $\sigma(I) \subseteq V$.* (b) *$H_1(I, V)$ is a closed submanifold of the Hilbert space $H_1(I, R^n)$.* (c) *If $p, q \in V$, then $\Omega(V, p, q)$ is a closed submanifold of $H_1(I, V)$.* (d) *If $\sigma \in H_1(I, V)$ then the tangent space to $H_1(I, V)$ at σ is $H_1(I, V)_\sigma = \{\lambda \in H_1(I, R^n) \mid \lambda(t) \in V_{\sigma(t)},\ t \in I\}$ and* (e) *if $\sigma \in \Omega(V, p, q)$ then the tangent space to $\Omega(V, p, q)$ at σ is just $\Omega(V; p, q)_\sigma = \{\lambda \in H_1(I, V)_\sigma | \lambda(0) = \lambda(1) = 0\}$.*

Proof: (a) is clear. It is equally clear that $H_1(I, V)$ is a closed set in $H_1(I, R^n)$ and that $\Omega(V, p, q)$ is a closed set in $H_1(I, V)$. Since V is a smooth submanifold of R^n we can find a smooth Riemann metric for R^n such that V is a totally geodesic submanifold. Then if $E : R^n \times R^n \to R^n$ is the corresponding exponential map (i.e. the map $t \to E(p, tv)$, where $E(p, tv)$ is the geodesic starting from p with tangent vector v), then E is a smooth map. Let $\sigma \in H_1(I, V)$ and define $\phi : H_1(I, R^n) \to H_1(I, R^n)$ by $\phi(\lambda)(t) = E(\sigma(t), \lambda(t))$. Then ϕ is smooth and $\phi(0) = \sigma$. Moreover $d\phi_0(\lambda)(t) = dE_0^{\sigma(t)}(\lambda(t))$, where $E^{\sigma(t)}(v) = E(\sigma(t), v)$. Since $dE_0^{\sigma(t)}$ is the identity map of R^n, $d\phi_0$ is the identity in $H_1(I, R^n)$. Thus by the inverse function theorem ϕ maps a neighborhood of zero in $H_1(I, R^n)$ C^k-isomorphically onto a neighborhood of σ in $H_1(I, R^n)$. Since V is totally geodesic, given λ near zero in $H_1(I, R^n)$, $\phi(\lambda) \in H_1(I, V)$ if and only if $\lambda \in H_1(I, V)_\sigma$. Similarly, if $\sigma \in \Omega(V, p, q)$ then $\phi(\lambda) \in \Omega(V, p, q)$ if and only if $\lambda \in \Omega(V, p, q)_\sigma$. So ϕ^{-1} restricted to a neighborhood of σ in $H_1(I, V)$ (resp. $\Omega(V, p, q)$) is a chart in $H_1(I, V)$ (resp. $\Omega(V, p, q)$) which is a restriction of a chart for $H_1(I, R^n)$. This completes the proof of (b) and (c), and the verification of (d) and (e) is routine.

Q.E.D.

Since any smooth map of a sub-manifold $V \subset R^n$ into a submanifold $W \subset R_m$ can be smoothly extended to a map from R^n to R^m, we can apply our results to obtain the following statement.

6.12. Theorem: *Let $V \subset R^n$, $W \subset R^m$ be smooth submanifolds, $\phi: V \to W$ a smooth map. Then $\bar{\phi}: H_1(I, V) \to H_1(I, W)$ defined by $\bar{\phi}(\sigma) = \phi\sigma$ is a smooth map of $H_1(I, V)$ into $H_1(I, W)$. Moreover the Frechet derivative*

$$d\bar{\phi}_\sigma : H_1(I, V)_\sigma \to H_1(I, W)_{\bar{\phi}(\sigma)}$$

is given by

$$d\bar{\phi}_\sigma(\lambda)(t) = d\phi_{\sigma(t)}(\lambda(t)).$$

Observe now that every manifold can, by Whitney's Theorem, be imbedded in an Euclidean space. Hence

6.13. Theorem: *$H_1(I, V)$ and $\Omega(V, p, q)$ are Hilbert manifolds, and, by Theorem* 6.12, *their manifold structure does not depend on the particular imbedding of V used.*

The function to which general Morse Theory will be applied in what follows will be the *action integral* $J^v(\sigma)$, defined for a Riemann manifold V as follows:

6.14. Definition: *For $\sigma \in H_1(I, V)$,*

$$J^v(\sigma) = \frac{1}{2}\int_0^1 |\sigma'(t)|^2 \, dt.$$

We leave to the reader proofs of the following properties of $J^v(\sigma)$.

6.15. Lemma: *Let V, W be smooth manifolds, $\phi: V \to W$ an isometry. Then $J^V = J^W \circ \phi$.*

6.16. Lemma: *Let V be a smooth submanifold of W. Then $J^V = J^W | H_1(I, V)$.*

6.17. Lemma: *J^V is a smooth functional.*

Advice: Prove Lemma 6.17 first for smooth submanifolds of R^n and then for general manifolds using Nash's imbedding theorem for Riemannian manifolds, which was proved as Theorem 2.4.

Next observe that $H_1(I, R^n)$, as a Hilbert space, has a natural Riemannian structure. Hence for all manifolds V, $H_1(I, V)$ will also have a

Riemannian structure. But the situation here is not so pleasant as in connection with the differentiable structure of $H_1(I, V)$, since now, in general, this Riemannian structure will depend on the imbedding $V \to R^n$. However, this will not bother us at all.

A second observation is the following. Let W be a complete Riemannian manifold, W_1 a closed submanifold of W inheriting from W its Riemannian structure; let ϱ_V, ϱ_W be the respective Riemannian metrics. It is clear that if $p, q \in W_1$, then $\varrho_V(p, q) \geqq \varrho_W(p, q)$, since, by definition the right side of given an infimum over a larger set than the infimum on the left. Hence the Riemannian structure of W_1 is also complete. Putting our observations together, we get

6.18. Theorem: *Let V be a smooth submanifold of R^n. Then $H_1(I, V)$ is a complete smooth Riemannian manifold with the Riemannian structure inherited from $H_1(I, R^n)$ where R^n is any Euclidean space in which V is isometrically imbedded.*

The same reasoning gives shows that $\Omega(V; p, q)$ is also a complete Riemannian manifold. Regarded as a submanifold of $H_1(I, R^n)$, the scalar product in it is simply $(\varrho, \lambda)_0 = (L\varrho, L\lambda)_0$.

One more preliminary needed for the application of Morse Theory is the verification of the Palais-Smale condition for the action integral. To remind the reader of the nature of this condition we write it down again:

P-S condition: Suppose that, for a sequence σ_n, $\nabla J^V(\sigma_n) \to 0$, and $J^V(\sigma_n)$ is bounded. Then there exists a subsequence σ_{n_k} convergent to an element $\sigma \in H_1(I, V)$.

We proceed to establish this condition for J^V in a series of substeps. We suppose throughout that V has been isometrically imbedded in a Euclidian space R^n. In what follows, L is the operator of Definition 6.3.

6.19. Lemma: *Let $\{\sigma_n\}$ be a sequence in $\Omega(V; p, q)$ such that $|L(\sigma_n - \sigma_m)|_0 \to 0$ as $n, m \to \infty$. Then σ_n converges in $\Omega(V; p, q)$.*

Proof: Evidently $\sigma_n - \sigma_m \in H_1^*(I, R^n)$. $\{\sigma_n\}$ is Cauchy in $H_1(I, R^n)$ and hence convergent. But $\Omega(V; p, q)$ is closed in $H_1(I, R^n)$.

Q.E.D.

6.20. Definition: *Let $p, q \in V$. If $\sigma \in \Omega(V; p, q)$ we define $h(\sigma)$ to be the orthogonal projection of $L\sigma$ onto the orthogonal complement of $L(\Omega(V; p, q)_\sigma)$ in $H_0(I, R^n)$.*

6.21. Theorem: *Let $J = J^V|\Omega(V; p, q)$. If we consider $\Omega(V; p, q)$ as a Riemannian manifold with the structure induced on it as a closed submanifold*

of $H_1(I, R^n)$ *then for each* $\sigma \in \Omega(V; p, q)$ $(\nabla J)(\sigma)$ *can be characterized as the unique element of* $\Omega(V; p, q)_\sigma$ *mapped by L onto* $L\sigma - h(\sigma)$. *Moreover* $|\nabla J(\sigma)|_\sigma = |L\sigma - h(\sigma)|_0$.

Proof: Note that $\Omega(V; p, q)_\sigma$ is a closed subspace of $H_1(I, R^n)$ and is contained in $H_1^*(I, R^n)$. It follows from Theorem 6.5 that L maps $\Omega(V; p, q)_\sigma$ isometrically onto a closed subspace of $H_0(I, R^n)$. Since $L\sigma - h(\sigma)$ is orthogonal to $L(\Omega(V; p, q)_\sigma)^\perp$, $L\sigma - h(\sigma) = L\lambda$, $\lambda \in \Omega(V; p, q)_\sigma$, with λ unique and $|\lambda|_\sigma = |L\lambda|_0 = |L\sigma - h(\sigma)|_0$. It will suffice to prove that $dJ_\sigma(\varrho) = (\lambda, \varrho)_\sigma$ for $\varrho \in \Omega(V; p, q)_\sigma$, i.e., that $dJ_\sigma(\varrho) = (L\lambda, L\varrho)_0 = (L\sigma - h(\sigma), L\varrho)_0$ for $\varrho \in \Omega(V; p, q)_\sigma$. Since $(h(\sigma), L\varrho)_0 = 0$ for $\varrho \in \Omega(V; p, q)_\sigma$, we must prove that $dJ_\sigma(\varrho) = (L\sigma, L\varrho)_0$ for $\varrho \in \Omega(V; p, q)_\sigma$. But $J^{R^n}(\alpha) = \frac{1}{2}|L\alpha|_0^2$, so $dJ_\sigma^{R^n}(\varrho) = (L\sigma, L\varrho)_0$ for $\varrho \in H_1(I, R^n)$. Since $J = J^{R^n}|\Omega(V; p, q)$, it follows that $dJ_\sigma = dJ_\sigma^{R^n}|\Omega(V; p, q)_\sigma$.

Q.E.D.

6.22. Definition: *Let* $\bar{\Omega}(V; p, q)_\sigma$ *be the closure of* $\Omega(V; p, q)_\sigma$ *in* $H_0(I, R^n)$, *and let* P_σ *be the orthogonal projection of* $H_0(I, R^n)$ *on* $\bar{\Omega}(V, p, q)_\sigma$. *For each point* $r \in V$, *let* $\Omega(r)$ *denote the orthogonal projection of* R^n *onto the tangent space* V_r *to* V *at* r.

6.23. Theorem: *The functional J of Theorem* 6.21 *satifies the Palais-Smale condition.*

Proof: Let $\{\sigma_n\}$ be a sequence in $H_1(I, V)$ such that $|J(\sigma_n)| \leqq M$, $\Delta J(\sigma_n) \to 0$. Since, by Theorem 6.21,

$$|\nabla J(\sigma_n)| = |L\sigma_n - h(\sigma_n)|_0,$$

we have $|L\sigma_n - h(\sigma_n)|_0 \to 0$. Since each P_σ is a projection—hence norm-decreasing—it follows from the corollary of Theorem 6.7 that $|L\sigma_n - P_{\sigma_n}h(\sigma_n)|_0 \to 0$, and by Corollary 2 of 6.7 we can assume on passing to a subsequence that $|\sigma_n - \sigma_m|_\infty \to 0$ as $m, n \to \infty$. We need only to prove that $|L(\sigma_n - \sigma_m)|_0 \to 0$ for $m, n \to \infty$, for then it follows that σ_n will converge in $\Omega(V; p, q)$ to a σ in $\Omega(V; p, q)$. But

$$|L(\sigma_n - \sigma_m)|_0^2 = (L\sigma_n, L(\sigma_n - \sigma_m))_0 - (L\sigma_m, L(\sigma_n - \sigma_m))_0.$$

Thus it suffices to prove that $(L\sigma_n, L(\sigma_n - \sigma_m))_0 \to 0$ as $m, n \to \infty$. Since $|L\sigma_n|^2 = 2J(\sigma_n)$ is bounded, $|L(\sigma_n - \sigma_m)|_0$ is bounded also and, since $L\sigma_n - P_{\sigma_n}h(\sigma_n) \to 0$ in $H_0(I, R^n)$ it suffices to prove that

$$(P_{\sigma_n}h(\sigma_n), L(\sigma_n - \sigma_m))_0 \to 0 \quad \text{as} \quad m, n \to \infty.$$

We now refer to Lemma 6.24 below and note that it follows from this Lemma that if $\sigma \in H_1(I, V)$ then $P_\sigma f$ belongs to $\Omega(V, p, q)_\sigma$ if f is smooth and vanishes for $t = 0$ and $t = 1$. Since $h(\sigma)$ is orthogonal to $LP_\sigma f$ in this case, we have $(h(\sigma), LP_\sigma f) = 0$ for all such σ and f. Thus

$$(*) \qquad (P_\sigma h(\sigma), Lf) = (h(\sigma), (P_\sigma L - LP_\sigma) f)$$

for $\sigma \in H_1(I, V)$ and smooth V vanishing at $t = 0{,}1$. If we put $Q_\sigma(t) = (d/dt)\, \Omega(\sigma(t))$, it follows by differentiation from (*) that

$$(P_\sigma h(\sigma), Lf) = (h(\sigma), Q_\sigma \cdot f) = (Q_\sigma \cdot h(\sigma), f)$$

for smooth f vanishing at $t = 0{,}1$, and hence, by a limit argument, for all $f \in H_1^*(I, R^n)$.

Since $\sigma_n - \sigma_m \in H_1^*(I, R^n)$ it follows that

$$|(P_{\sigma_n} h(\sigma_n), L(\sigma_n - \sigma_m))_0| = \left| \int_0^1 (Q_{\sigma_n}(t)\, h(\sigma_n)(t), (\sigma_n - \sigma_m)(t))\, dt \right|$$

$$\leqq |\sigma_n - \sigma_m|_\infty \int_0^1 |Q_{\sigma_n}(t)\, h(\sigma_n)(t)|\, dt$$

is bounded. Let A be a compact set such that $\sigma_n(I) \subset A$. Then there exists K such that

$$\int_0^1 |Q_{\sigma_n}(t)\, h(\sigma_n)(t)|\, dt \leqq K\, |L\sigma_n|_0\, |h(\sigma_n)|_0 .$$

Now, since $|L\sigma_n|_0$ is bounded and since $|L\sigma_n - h(\sigma_n)|_0 \to 0$, $|h(\sigma_n)|_0$ is bounded and the theorem follows.

Q.E.D.

Finally, we relate critical points of the action integral J with geodesics, and find conditions under which these critical points are non-degenerate. We will not discuss the geometry of geodesics of a finite-dimensional manifold in detail, but refer the reader instead to (3) or (4) of the Bibliography.

6.24. Lemma: *Let $\sigma \in \Omega(V; p, q)$. Then $\bar{\Omega}(V; p, q)_\sigma = \{\lambda \in H_0(I, R^n) \mid \lambda(t) \in V_{\sigma(t)}$ for almost all $t \in I\}$. If $\lambda \in H_0(I, R^n)$ then $(P_\sigma \lambda)(t) = \Omega(\sigma(t))\, \lambda(t)$.*

Proof: Let $\pi_\sigma \in L(H_0(I, R^n), H_0(I, R^n))$ be defined by $(\pi_\sigma \lambda)(t) = \Omega(\sigma(t))\, \lambda(t)$. Since $\Omega(\sigma(t))$ is an orthogonal projection in R^n for each $t \in I$ it follows from the definition of the inner product in $H_0(I, R^n)$ that π_σ is an orthogonal projection. From the characterization of $\Omega(V; p, q)_\sigma$ it is clear that π_σ maps $H_1^*(I, R^n)$ onto $\Omega(V; p, q)_\sigma$. Since $H_1^*(I, R^n)$ is dense in $H_0(I, R^n)$

it follows that the range of π_σ is $\bar{\Omega}(V; p, q)_\sigma$, so $\pi_\sigma = P_\sigma$. On the other hand, $\lambda \in H_0(I, R^n)$ is fixed under π_σ if and only if $\lambda(t) \in V_{\sigma(t)}$ for almost all $t \in I$. Since the range of a projection is its set of fixed points, this proves our lemma.

Q.E.D.

The following are obvious consequences of the lemma.

Corollary 1: *If $\sigma \in \Omega(V; p, q)$, then*

$$P_\sigma(H_1(I, R^n)) = H_1(I, V)_\sigma \quad \text{and} \quad P_\sigma(H_1^*(I, R^n)) = \Omega(V; p, q)_\sigma$$

Corollary 2: *If $\sigma \in \Omega(V; p, q)$ then $P_\sigma L\sigma = L\sigma$.*

Another simple result, whose proof is left as an exercise, is

6.25. Lemma: *Let $T \in H_0(I, L(R^n, R^p))$ and define for each $\lambda \in H_0(I, R^n)$ a measurable function $\bar{T}(\lambda): I \to R^p$ by $\bar{T}(\lambda)(t) = T(t)\lambda(t)$. Then*

(1) *$\bar{T}$ is bounded from $H_0(I, R^n)$ to $L^1(I, R^p)$;*

(2) *If T and λ are absolutely continuous then so is $\bar{T}(\lambda)$ and*

$$(\bar{T}\lambda)'(t) = T'(t)\lambda(t) + T(t)\lambda'(t);$$

(3) *If $\bar{T} \in H_1(I, L(R^n, R^p))$, $\lambda \in H_1(I, R^n)$, then $\bar{T}(\lambda) \in H_1(I, R^p)$.*

6.26. Definition: Let $\sigma \in \Omega(V; p, q)$. Define $G_\sigma \in H_1(I, L(R^n, R^n))$ by $G_\sigma = \Omega \circ \sigma$ and $Q_\sigma \in H_0(I, L(R^n, R^n))$ by $Q_\sigma = G'_\sigma$.

6.27. Theorem: *Let $\sigma \in \Omega(V; p, q)$. Let P_σ be as in Definition 6.22. If $\varrho \in H_1(I, R^n)$, then $(LP_\sigma - P_\sigma L)\varrho(t) = Q_\sigma(t)\varrho(t)$. Given $f \in H_0(I, R^n)$, define an absolutely continuous map $g: I \to R^n$ by*

$$g(t) = \int_0^t Q_\sigma(s) f(s)\, ds.$$

Then, if $\varrho \in H_1^*(I, R^n)$

$$(f, (LP_\sigma - P_\sigma L)\varrho)_0 = (g, -L\varrho)_0.$$

Proof: Since $P_\sigma\varrho(t) = G_\sigma(t)\varrho(t)$ and $P_\sigma(L\varrho)(t) = G_\sigma(t)\varrho'(t)$ by 6.24, $(LP_\sigma - P_\sigma L)(\varrho(t)) = Q_\sigma(t)\varrho(t)$ follows immediately by differentiation. By (1) of Lemma 6.25, $s \to Q_\sigma(s) f(s)$ is summable, so g is absolutely continuous. Next note that, since $G_\sigma(t) = \Omega(\sigma(t))$ is self-adjoint for all t, $Q_\sigma(t) = G'_\sigma(t)$ is self-adjoint wherever defined, and hence

$$(f, (LP_\sigma - P_\sigma L)\varrho)_0 = \int_0^1 (f(t), Q_\sigma(t)\varrho(t))\, dt = \int_0^1 (Q_\sigma(t) f(t), \varrho(t))\, dt$$

$$= \int_0^1 (g'(t), \varrho(t))\, dt.$$

Then if $\varrho \in H_1^*(I, R^n)$ Theorem 6.5 gives

$$(f, (LP_\sigma - P_\sigma L)\varrho)_0 = (g, -L\varrho)_0 .$$

Q.E.D.

6.28. Theorem: *Let $h(\sigma)$ be as in Definition 6.20. If $\sigma \in \Omega(V; p, q)$ then $P_\sigma h(\sigma)$ is absolutely continuous and $(P_\sigma h(\sigma))'(t) = Q_\sigma(t)\, h(\sigma)(t)$.*

Proof: If $\varrho \in H_1^*(I, R^n)$ then

$$(P_\sigma h(\sigma), L\varrho)_0 = (h(\sigma), P_\sigma L\varrho)_0 = (h(\sigma), (P_\sigma L - LP_\sigma)\varrho)_0$$

since $(h(\sigma), LP_\sigma\varrho) = 0$. Hence $(P_\sigma h(\sigma), L\varrho)_0 = (g, L\varrho)_0$ if we define g to be

$$g(t) = \int_0^t Q_\sigma(s)\, h(\sigma)(s)\, ds .$$

Then $P_\sigma h(\sigma) - g \perp L(H_1^*(I, R^n))$, whence $P_\sigma h(\sigma) - g =$ constant. Since g is absolutely continuous so is $P_\sigma h(\sigma)$ and they have the same derivative. But $g'(t) = Q_\sigma(t)\, h(\sigma)(t)$.

Q.E.D.

6.29. Theorem: *Let σ be a critical point of J. Then σ is smooth and, moreover $\sigma'' \perp V$ everywhere. Conversely, if $\sigma \in \Omega(V; p, q)$, σ' a.e., $\sigma'' \perp V$, then σ is a critical point of J.*

Proof: By Theorem 6.21, if σ is a critical point of J, then $L\sigma = h(\sigma)$. Since $P_\sigma L\sigma = L\sigma$, it follows that $P_\sigma h(\sigma) = h(\sigma)$, so by Theorem 6.21 σ' is absolutely continuous (so that σ is C^1) and

(*) $$\sigma''(t) = Q_\sigma(t)\, \sigma'(t).$$

Now since $\Omega : V \to L(R^n, R^n)$ is smooth using 6.26 we have

$$Q_\sigma(t) = \frac{d}{dt}\, \Omega(\sigma(t)) .$$

It follows that if σ is C^m, then $Q_\sigma(t)$ is C^{m-1}, so by (*) the statement that σ'' is C^{m-1} implies that σ is C^{m+1}. Since we already know σ is C^1, it follows that σ is smooth. If $\varrho \in \Omega(V; p, q)_\sigma$, then $L\sigma = h(\sigma)$ is orthogonal to $L\varrho$, so that σ'' is orthogonal to ϱ. Since σ'' and ϱ are continuous, it follows that $(\sigma''(t), \varrho(t)) = 0$, $t \in I$. If $t \in I$ is not an endpoint and $v_0 \in V_{\sigma(t)}$, then there exists $\varrho \in \Omega(V; p, q)$ such that $\varrho(t) = v_0$, hence $\sigma''(t)$ is orthogonal to $V_{\sigma(t)}$ and, by continuity, this holds also at the endpoints. Conversely, if $\sigma \in \Omega(V; p, q)$ is such that σ' is absolutely continuous and $\sigma'' \perp V_{\sigma(t)}$ for almost all $t \in I$, then $L\sigma \perp L(\Omega(V; p, q))$ so $L\sigma = h(\sigma)$ and σ is a critical point of J.

Q.E.D.

The last step in the characterization of critical points of J is supplied by the well-known result of classical differential geometry (see (3) and (4)) that, if $\sigma \in C^2(I, V)$, σ is a geodesic of V parametrized proportionately to arc-length if and only if $\sigma'' \perp V$ everywhere. We obtain the following conclusion.

6.30. Theorem: *If $\sigma \in \Omega(V; p, q)$, then σ is a critical point of J if and only if σ is a geodesic of V parametrized proportionately to arc length.*

We must now determine when an extremal point of J will be degenerate. We limit ourselves to a brief exposition and to suggesting that the reader consult (3).

Let E denote the exponential map of V_p into V; i.e. if $v \in V_p$, then $E(v) = \sigma(|v|)$ where σ is the geodesic starting from p with tangent vector $v/|v|$. Then E is smooth. Given $v \in V_p$ we define $\lambda(v) =$ dimension of null-space of dE_v. If $\lambda(v) > 0$, we call v a *conjugate vector at p*. A point of V is called a *conjugate point* of p if it is in the image under E of the set of conjugate vectors at p. By Sard's theorem the set of conjugate points of p has measure zero in V.

Given $v \in E^{-1}(p)$ define $\tilde{v} \in \Omega(V; p, q)$ by $\tilde{v}(t) = E(t(v))$. Then $\tilde{v}$ is a geodesic parametrized proportionately to arc length (factor: $|v|$) and hence a critical point of J. Conversely, any critical point of J is of the form $\tilde{v}$ for a unique $v \in E^{-1}(q)$. We may now state the following two theorems:

6.30. Non-degeneracy theorem: *If $v \in E^{-1}(q)$ then $\tilde{v}$ is a degenerate critical point of J if and only if v is a conjugate vector at p. Hence J has only non-degnerate critical points if and only if q is not a conjugate point p. This condition is satisfied if q lies outside of a set of measure zero in V.*

6.31. Morse index theorem: *Let $v \in E^{-1}(q)$. Then there are only a finite number of t satisfying $0 < t < 1$ such that t_v is a conjugate vector at p. The index of $\tilde{v}$ is $\sum_{0<t<1} \lambda(tv)$. In particular each critical point of J has finite index.*

Proof of 6.31 and 6.32

Let σ be a geodesic on the manifold V. By definition 6.14 of $J^V(\sigma)$ we have

$$J^V(\sigma) = \frac{1}{2}\int_0^1 \frac{d\sigma}{dt}(t) \cdot \frac{d\sigma}{dt}(t)\, dt,$$

so that, if we introduce coordinates in the neighborhood of the curve σ we have

$$J^V(\sigma) = \frac{1}{2}\int_0^1 g_{ij}(\sigma(t)) \frac{d\sigma^i}{dt}(t) \frac{d\sigma^j}{dt}(t)\, dt.$$

If we then put $\sigma_\varepsilon = \sigma + \varepsilon\varrho$, and calculate the terms of second order in ε in order to evaluate the Hessian quadratic form $\delta^2 J^v(\sigma; \varrho, \varrho)$ we find that

$$\delta^2 J^V(\sigma; \varrho, \varrho)$$
$$= \frac{1}{2}\int_0^1 \left\{ g_{ij}(\sigma(t)) \frac{d\varrho^i(t)}{dt}\frac{d\varrho^j(t)}{dt} + A_{ij}(t)\,\varrho^i(t)\frac{d\varrho^j(t)}{dt} + B_{ij}(t)\,\varrho^i(t)\,\varrho^j(t) \right\} dt,$$

where $A_{ij}(t)$ and $B_{ij}(t)$ may readily be expressed in terms of σ and of the first and second partial derivatives of g_{ij}.

If we are careful to choose coordinates in the neighborhood of the geodesic curve σ in such a way that σ itself is the first axis and curves perpendicular to σ give the remaining axes, we have $g_{ij}(\sigma(t)) = \delta_{ij}$, and the above expression for the Hessian reduces to

[†] $\delta^2 J^V(\sigma; \varrho, \varrho)$

$$= \frac{1}{2}\int_0^1 \left\{ \sum_{i=1}^n \left(\frac{d\varrho^i(t)}{dt}\right)^2 + A_{ij}(t)\,\varrho^i(t)\frac{d\varrho^j(t)}{dt} + B_{ij}(t)\,\varrho^i(t)\,\varrho^j(t) \right\} dt.$$

The Hessian matrix $\delta^2 J(\sigma, \ldots, .)$ will be singular if and only if there exists a function $\varrho \in \Omega(V; p, q)_\sigma$ such that $\delta^2 J(\sigma; \varrho, \tilde{\varrho}) = 0$ for all $\tilde{\varrho} \in \Omega(V; p, q)_\sigma$. That is, the Hessian matrix will be singular if and only if there exists a non-zero function $(\varrho^i) \in H_1(I)$ such that $\varrho^i(0) = \varrho^i(1) = 0$ and such that

$$Q_\sigma(\varrho, \tilde{\varrho}) = \int_0^1 \left\{ \sum_{i=1}^n \frac{d\varrho^i(t)}{dt}\frac{d\tilde{\varrho}(t)}{dt} + \frac{1}{2} A_{ij}(t)\,\varrho^i(t)\frac{d\tilde{\varrho}^j(t)}{dt} \right.$$
$$\left. + \frac{1}{2} A_{ij}(t)\,\tilde{\varrho}^i(t)\frac{d\varrho^j(t)}{dt} + B_{ij}(t)\,\tilde{\varrho}^i(t)\,\varrho^j(t) \right\} dt = 0$$

for all $(\tilde{\varrho}^i) \in H_1([0, 1])$ such that $\tilde{\varrho}^i(0) = \tilde{\varrho}^i(1) = 0$.

Integrating the above expression by parts, we see that $\sigma^2 J(\sigma; \cdot, \cdot)$ will be singular if and only if the second order differential system

$$(*) \qquad -\frac{d^2\varrho^i(t)}{dt^2} + \frac{1}{2} A_{ij}(t)\frac{d\varrho^j(t)}{dt} - \frac{1}{2}\frac{d}{dt}\{A_{ji}(t)\,\varrho^j(t)\} + B_{ij}(t)\,\varrho^j(t) = 0$$

has a solution (ϱ^i) satisfying $\lambda = 0$, $\varrho^i(0) = 0$, $\varrho^i(1) = 0$.

Call the real numbers λ for which there exists a non-zero function (ϱ^i) with $\varrho^i(0) = 0 = \varrho^i(1)$ satisfying (*) the *eigenvalues* of the Hessian form Q_σ; call the corresponding functions ϱ the *eigenfunctions* belonging to the eigenvalue λ; and call the number of linearly independent functions belonging to the eigenvalue λ the *multiplicity of the eigenvalue* λ.

Then the classical theory of Sturm-Liouville equations supplies the following results.

a. Every eigenvalue of the boundary value problem defined by the differential equation (*) and the boundary conditions $\varrho^i(0) = \varrho^i(1) = 0$ is of finite multiplicity. The eigenvalues form an infinite sequence of isolated points bounded below. Thus, if the eigenvalues are enumerated in increasing order, each being repeated a number of times equal to its multiplicity, they form an increasing sequence $\lambda_1, \lambda_2, \ldots,$ of real numbers approaching infinity.

b. (Minimax principle). Let $H_1^{(0)}([0, 1])$ denote the set of functions $\varrho = (\varrho^i) \in H_1([0, 1])$ such that $\varrho^i(0) = 0 = \varrho^i(1)$. Put

$$\varrho\,(\tilde{\varrho}) = \sum_{i=1}^{n} \int_0^1 \varrho^i(t)\, \tilde{\varrho}^i(t)\, dt,$$

and $|\varrho| = (\varrho, \varrho)^{1/2}$. Then the k-th eigenvalue λ_k in the above sequence is given by the expression

$$(**) \qquad \lambda_k = \max_{\substack{\sigma_1, \ldots, \sigma_{k-1} \\ |\varrho| = 1}} \quad \min_{\substack{(\varrho, \sigma_j) = 0,\, j = 1, \ldots, k-1 \\ \varrho \in H_1^{(0)}([0, 1])}} Q_\sigma(\varrho, \varrho)$$

Let $0 \leqq a \leqq 1$, and let $H_1([0, a])$ denote the set of all functions $\varrho = (\varrho^i)$ on the interval $[0, a]$ which have square-integrable first derivatives: let $H_1^{(0)}([0, a])$ denote all those which vanish at both end-points of the interval $[0, a]$. Let $\lambda_n(a)$ be the n-th eigenvalue, in increasing order and with repetitions according to multiplicity, of the equation (*), with boundary conditions $\varrho^i(0) = 0 = \varrho^i(a)$. Then, applying the minimax principle (**), we find that

$$(***) \qquad \lambda_k(a) = \max_{\substack{\sigma_1, \ldots, \sigma_{k-1} \\ \sigma_i \in H_1^{(0)}([0, a]) \\ |\varrho| = 1}} \quad \min_{\substack{(\varrho, \sigma_j) = 0 \\ j = 1, \ldots, k-1 \\ \varrho \in H_1^{(0)}([0, a])}} Q(\varrho, \varrho).$$

Since $H_1^{(0)}([0, a])$ may also be regarded as the subset of $H_1^{(0)}([0, 1])$ consisting of all functions $\varrho = (\varrho^i)$ such that $\varrho^i(t) = 0, a \leqq t \leqq 1$, it follows immediately, on comparing (**) and (***), that $\lambda_k(a) \geqq \lambda_k(1) = \lambda_k$ for all k. By the same argument but more generally we have $\lambda_k(a) \geqq \lambda_k(b)$ for $a \leqq b$. Thus the eigenvalues $\lambda_k(a)$, regarded as functions of the parameter a, are monotone-decreasing. For all a we have $\varrho^i(t) = \int_0^t (\varrho^i(s))'\, ds \leqq t^{1/2} \left\{ \int_0^1 |(\varrho^i(s))'|^2\, ds \right\}^{1/2}$; thus for all sufficiently small values of a the first term of the integral [†] dominates the others, and the expression [†] is necessarily positive. By the minimax principle (**), this implies that for sufficiently small a, all the eigenvalues $\lambda_k(a)$ are positive. Thus, for each $a > 0$, the number of negative eigenvalues is precisely equal to the number of eigenvalues which have crossed from positive to negative as a has increased from zero to its given value.

The above arguments establish the following lemma:

6.32. Lemma:

(i) The Hessian matrix $\delta^2 J^v(\sigma; \varrho, \tilde{\varrho})$ is singular, i.e., the critical point σ of the functional J^V is degenerate, if and only if the equation (*) has a non-zero solution $\varrho = (\varrho^i)$ satisfying $\varrho^i(0) = \varrho^i(1) = 0$.

(ii) Let $0 < a_1 < a_2 < \cdots < a_p < 1$ be the values of a for which the differential equation (*) has a non-zero solution $\varrho = (\varrho^i)$ satisfying $\varrho^i(0) = \varrho^i(a) = 0$; and let $n(a)$ be the number of such linearly independent solutions attaching to the value a. Then Morse index of the critical point σ, i.e., the number of negative eigenvalues of the hessian matrix $\delta^2 J^V$, is equal to the sum $n(a_1) + \cdots + n(a_p)$.

Next we shall need the following Lemma.

6.33. Lemma: Let $n(a)$ be defined for each a as in (ii) of the preceeding Lemma. Let $v \to E(v)$ be the exponential transformation, which sends each tangent vector v at the point ϱ of the manifold V into the point $\sigma_v(|v|)$, where σ_v is the geodesic starting from P with tangent vector $v/|v|$. Let v_0 be such that $\sigma = \sigma_{v_0}$. Let dE_{v_0} be the gradient of the map E at the point v_0. Then $n(1)$ is equal to the dimension of the null-space of the linear transformation dE_{v_0}.

Before giving the proof of this Lemma, let us note that the Non-degeneracy Theorem and the Morse Index Theorem follow readily from the two preceeding lemmas. Indeed, since $n(1) = 0$ is the criteria for non-degeneracy according to the first of our two lemmas, the Non-degeneracy Theorem follows immediately from the second lemma. As to the Morse Index Theorem, we note that, applying the second of Lemmas to each of the geodesic segments $\sigma(t)$, $0 \leqq t \leqq a$, with $a \leqq 1$, we find that $n(a) = \lambda(av_0)$ for $0 \leqq a \leqq 1$. Thus the Morse Index Theorem follows at once from part (ii) of the first of our Lemmas.

Let us now give the proof of the second lemma:

Proof: Put $\sigma_v(t) = E(vt)$, so that σ_v is a geodesic curve parametrized proportional to arc-length, whose tangent vector at $t = 0$ is v. Since for any v σ_v is a critical point of the functional $J^V(\sigma)$ we have $\frac{d}{d\varepsilon} J^V(\sigma_v + \varepsilon\varrho)|_{\varepsilon=0} = 0$ for every function $\varrho = (\varrho^i)$ vanishing for $t = 0$ and $t = 1$. Thus, if $\bar{v}$ is any vector tangent to V at the same point p as v, we have

$$\frac{\partial^2}{\partial\varepsilon\partial\bar{\varepsilon}} J^V(\sigma_{v+\bar{\varepsilon}\bar{v}} + \varepsilon\varrho)\Big|_{\substack{\varepsilon=0\\ \bar{\varepsilon}=0}} = 0$$

for all ϱ. That is,

$$\delta^2 J^V\left(\sigma_v; \frac{d}{d\bar{\varepsilon}}\sigma_{v+\bar{\varepsilon}\bar{v}}, \varrho\right)\Big|_{\substack{\varepsilon=0\\ \bar{\varepsilon}=0}} = 0$$

for all ϱ vanishing for $t = 0$ and $t = 1$. If we note that the differential equation [*] is derived from the variational condition [†] on integrating by parts, we see at once from this last equation that the function

$$\Delta_{\bar{v}}(t) = \frac{d}{d\bar{\varepsilon}}\sigma_{v+\bar{v}\bar{\varepsilon}}(t)\Big|_{\bar{\varepsilon}=0}$$

satisfies the linear differential equation [*].

We have

$$\frac{d}{dt}\Delta_{\bar{v}}(t)\Big|_{t=0} = \frac{\partial^2}{\partial\bar{\varepsilon}\partial t}\sigma_{v+\bar{\varepsilon}\bar{v}}(t)\Big|_{t=\varepsilon=0} = \frac{d}{d\bar{\varepsilon}}(v + \bar{\varepsilon}\bar{v})\Big|_{\bar{\varepsilon}} = 0 = \bar{v};$$

thus $\Delta_{\bar{v}}$ satisfies the initial conditions $\Delta_{\bar{v}}(0) = 0$, $\Delta'_{\bar{v}}(0) = \bar{v}$. On the other hand, taking $t = 1$, we find that $\Delta_{\bar{v}}(1) = \frac{d}{d\bar{\varepsilon}}\sigma_{v+\bar{\varepsilon}\bar{v}}(t)\Big|_{\varepsilon=0} = \frac{d}{d\bar{\varepsilon}}E(v + \bar{\varepsilon}\bar{v})\Big|_{\bar{\varepsilon}=0}$ $= dE_v(\bar{v})$. Thus the dimension of the null-space of dE_v is at the same time the dimension of the null-space of $\Delta_{\bar{v}}(1)$; and hence equals the dimension of the space of vectors $\bar{v}$ such that $\Delta_{\bar{v}}$ satisfies both the boundary conditions $\Delta_{\bar{v}}(0) = 0$ and $\Delta_{\bar{v}}(1) = 0$. That is, the dimension of the null-space of dE_v is the integer $n(1)$ of Lemma 6.33, and thus the proof of Lemma 6.33 is complete.

Q.E.D.

CHAPTER VII

Applications

A. Applications to Homotopy Theory

We first recall the definition of the homotopy groups of a space. Let X be a topological space, A a subspace of X and $p \in A$. Let I^n denote the n-dimensional cube, $I^{n-1} \subset I^n$ the bottom face, and J^{n-1} the union of all the other faces, so that $J^{n-1} = \partial I^n - I^{n-1}$. We shall write

$$f: (I^n, I^{n-1}, J^{n-1}) \to (X, A, p)$$

for any continuous function $f: I^n \to X$ which maps I^{n-1} into A and J^{n-1} on p. We denote by $\Omega_n(X, A, p)$ the space of all such functions, and by $\pi_n(X, A, p)$ the set of all components of $\Omega_n(X, A, p)$. $\pi_n(X, A, p)$ has a well known group structure (by taking representatives of two elements of π_n, reparametrizing and then "joining" them). We call it the n-dimensional homotopy group of X relative to A with base point p.

The following are easily proved properties of the groups π_n:

(i) By reparametrizing we get for $n \geqq 2$

$$\Omega_n(X, A, p) \cong \Omega_1(\Omega_{n-1}(X, A, p), \phi_0, \phi_0),$$

where ϕ_0 is the constant map sending I^{n-1} to p. Hence, $\pi_n(X, A, p) = \pi_1(\Omega_{n-1}(X, A, p), \phi_0, \phi_0)$.

(ii) In the same way we prove that for $n \geqq 2$

$$\pi_n(X, A, p) = \pi_{n-1}(\Omega(X, A, p), \phi_0, \phi_0),$$

where ϕ_0 sends I^1 onto p.

(iii) The identity of $\pi_n(X, A, p)$ is the set of maps homotopic to the constant map $\phi_0 : I^n \to p$.

(iv) If $\phi(I^n) \subset A$, then ϕ is homotopic to the identity.

(Shrink I^n by means of a function m_t, so that $\phi_t = \phi m_t$ is a homotopy between ϕ and ϕ_0.)

In other words, $\pi_n(X, X, p)$ is trivial for any $p \in X$.

In the sequel we shall write $\pi_n(X, p)$ for $\pi_n(X, p, p)$. It is easy to see that $\pi_n(X, p)$ is, in fact, the usual "absolute" n-dimensional homotopy group of X with base point p. Also, $\pi_n(X, A, p)$ will often be written $\pi_n(X, A)$, when no confusion can arise. (Of course if A is arc-wise connected $\pi_n(X, A, p)$ does not depend on p.)

Suppose we have a map $\psi : (X, A, p) \to (Y, B, q)$. Then ψ induces a map $\psi^* : \pi_n(X, A, p) \to \pi_n(Y, B, q)$. (Just send $\phi \in \Omega_n(X, A, p)$ into $\psi\phi \in \Omega_n(Y, B, q)$.)

7.1. Definition: The *boundary homomorphism* $\partial : \pi_n(X, A, p) \to \pi_{n-1}(A, p, p)$ or briefly $\partial : \pi_n(X, A) \to \pi_{n-1}(A)$ is defined as follows.

Given $q \in \pi_n(X, A)$, take $\phi \in q$, then $\phi | I^{n-1}$ belongs to $\Omega_{n-1}(A, p, p)$ and so determines a class $\partial q \in \pi_{n-1}(A)$.

We state the following without proof.

7.2. Theorem: *Let i be the injection $(A, p) \to (X, p)$ and j the injection $(X, p) \to (X, A)$. Then the following sequence is exact:*

$$\cdots \longrightarrow \pi_n(X, A) \xrightarrow{\partial} \pi_{n-1}(A, p) \xrightarrow{i^*} \pi_{n-1}(X, p) \xrightarrow{j^*} \pi_{n-1}(X, A) \longrightarrow \cdots$$

This is the analog for homotopy groups of the Exactness Principle given in Chapter IV, Part 2, § E, of these Notes.

Now, suppose we have a manifold $M, f \in C^\infty(M)$, satisfying the P-S condition, and let as usual $M^a = \{x \in M; f(x) \leqq a\}$, $M^b = \{x \in M; f(x) \leqq b\}$. If there are only non-degenerate critical levels between a and b, M^b is deformable to M^a with handles attached:

$$(1) \qquad M^b \sim M^a \underset{h_1}{\cup} h_1(D^{k_1} \times D^{l_1}) \underset{h_2}{\cup} h_2(D^{k_2} \times D^{l_2}) \underset{h_3}{\cup} \cdots$$

Let Λ be another manifold, and ϕ a mapping $\phi : \Lambda \to M^b$. Assume that $\dim(\Lambda)$ is less than the index k_j of any critical point in (1) and that Λ is compact. Next, note that ϕ_l can be deformed to a smooth map, and set $\Lambda_1 = \phi^{-1}(h_1(D^{k_1} \times D^{l_1}))$. Consider $h_1^{-1}\phi : \Lambda_1 \to D^{k_1} \times D^{l_1}$. Let p_1 be the projection map of $D^{k_1} \times D^{l_1}$ onto D^{k_1}. Then $p_1 h_1^{-1}\phi$ is smooth and maps Λ_1 into D^{k_1}. But $\dim(\Lambda_1) < k_1$, whence some point in D^{k_1} does not belong to the range of $p_1 h_1^{-1}\phi$, and the same holds for the other indices k_2, etc.

Now a manifold of the type

$$M^a \underset{h_1}{\cup} h_1 (D^{k_1} - q_1 \times D^{l_1}) \underset{h_2}{\cup} h_2 (D^{k_2} - q_2 \times D^{l_2}) \underset{h_3}{\cup} \cdots$$

can be deformed into M^a (see drawing).

Hence, ϕ can be deformed to a map $\Lambda \to M^a$.

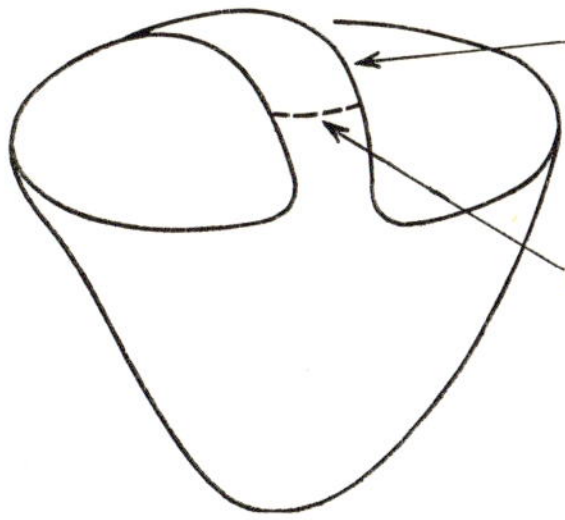

As a special case we get:

7.3. Theorem: $\pi_n (M^b, M^a) = 0$ if $n <$ degree of any critical point between a and b.

7.4. Corollary: If Morse theory applies to (M, f) and if above some non-critical level c all critical points have indices greater than n, then $\pi_n (M, M^c) = 0$.

We will now apply our results to the topology of spheres, in order to obtain the so-called Freudenthal suspension relation between homotopy groups.

First we recall that in relation to $H_1(S^j, p, q)$ and the function J, the geodesics joining p and q are critical points whose indices depend on the length of the geodesic: if length $(\gamma) = \pi - \varepsilon$ for any $0 < \varepsilon < \pi$, then index $(\gamma) = 0$; if length $(\gamma) = \pi + \varepsilon$, then index $(\gamma) = j - 1$; if length $(\gamma) = 3\pi - \varepsilon$, index $(\gamma) = 2(j - 1)$ and so on.

This follows from the Morse Index Theorem 6.31. Suppose that we have, as before, a map $\phi: \Lambda \to H_1(S^j; p, q)$ where $p \neq q$ and $p \neq q'$, the conjugate of q, and that $\dim (\Lambda) < 2(j - 1)$. Then by 7.3 ϕ is homotopic to a map whose range contains curves of length at most $\pi + 2\varepsilon$. Now assume that length $(\sigma) < \pi + 2\varepsilon$. Let m be the midpoint of σ: $m = \sigma(t_*)$. It is easy to see that $m: H_1(S^j; p, q) \to S^j$ is a smooth map.

We have $d(p, m) < \frac{1}{2}\pi + \varepsilon$ and $d(q, m) < \frac{1}{2}\pi + \varepsilon$. This implies that there are unique geodesics σ_1 joining p and m and σ_2 joining q and m (see drawing below), if ε is small enough.

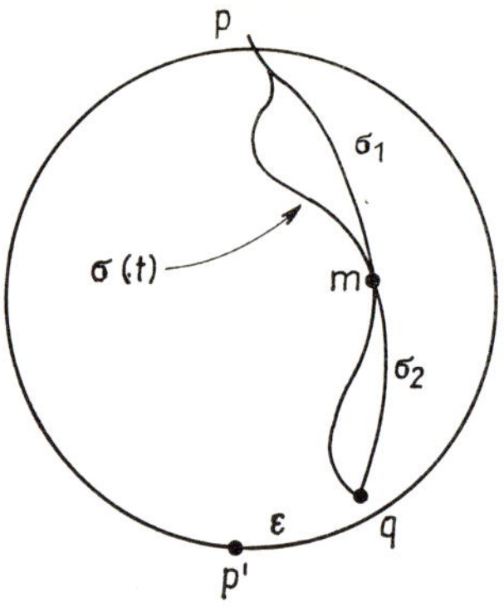

Then $d(\sigma(t), \sigma_1(t)) \leqq \frac{1}{2}(\pi + 2\varepsilon) + \frac{1}{4}(\pi + 2\varepsilon) < \pi$ if ε is small enough (ε being the distance between p' and q). Hence in this case $\sigma(t)$ and $\sigma_1(t)$ are connected by a unique shortest geodesic varying continuously with t, whence σ can be deformed through these geodesics into σ_1. The same holds for σ_2. Thus any map $\phi : \Lambda \to H_1(S^j; p, q)$ is homotopic to a map $\phi : \Lambda \to H_1(S^j; p, q)$ such that each value $\phi(\Lambda)$ is a broken geodesic of two segments and total length less than $\pi + 2\varepsilon$. It follows that the space of maps $\Lambda \to H_1$ is of the same homotopy type as the space of maps with values in a "belt", and hence of the same homotopy type as the space of maps $\Lambda \to S^{j-1}$ (see figure below).

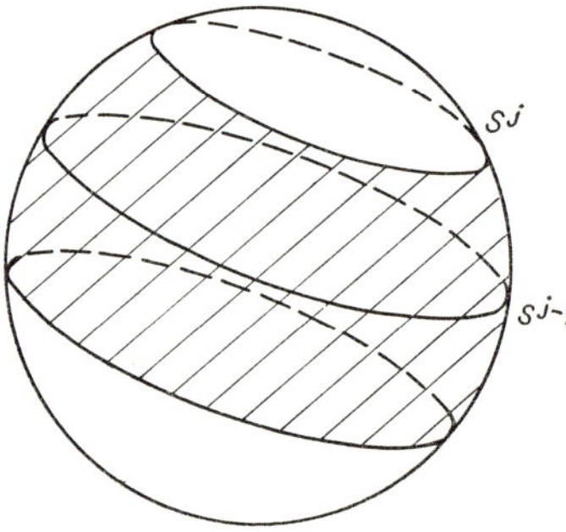

Now, it can readily be proved that $H_1(S^j; p, q)$ is of the same homotopy type as $H_1(S^j; \check{p}, \check{q})$; that is, the homotopy type does not depend on the points p and q. Thus our result is independent of the relative position of p and q.

In particular, we obtain:

7.5. Theorem: If $\dim(\Lambda) < 2(j-1)$, the space of maps $\Lambda \to \Omega_1(S^j; p, q)$ is of the same homotopy type as the space of maps $\Lambda \to S^{j-1}$.

7.6. Corollary: For $n < 2(j-1)$

$$\pi_n(\Omega_1(S^j; p, q)) \simeq \pi_n(S^{j-1}).$$

By property (ii) of the homotopy groups, we obtain

7.7. Corollary:

$$\pi_{n+1}(S^{j+1}) \simeq \pi_n(S^j), \quad \text{for} \quad n < 2j.$$

Corollary 7.7 is known as the *Freudenthal suspension relation.*

7.8. Corollary:

$$\pi_n(S^n) \simeq \pi_{n+1}(S^{n+1}), \quad \text{if} \quad n > 0,$$

whence $\pi_n(S^n) = Z$ if $n > 0$.

B. A Proof of Theorem 5.16

Let $X \xrightarrow{p} B$ be a fiber space. Also let $\phi : A \to X$ and $\psi = p\phi : A \to B$. We say that the homotopy ψ_t of ψ has the "lifting property" if there exists a homotopy ϕ_t of ϕ such that $\psi_t = p\phi_t$.

Example: If $X = B \times C$ and $p_1 : X \to B$ is the natural projection on B and $p_2 : X \to C$ that on C, and if ϕ and ψ are two functions as above, then given a homotopy ψ_t the map

$$\phi_t = \psi_t \times p_2\phi : A \to X$$

has the required properties.

We state without proof the following

7.9. Theorem (Künneth) [Cf., for example, Hilton and Wiley, *Homology Theory.*]

$$H_n(B \times C; G) = \sum_{k+l=n} \oplus H_k(B; H_l(C; G)).$$

7.10. Corollary: Suppose G = real numbers. Then

$$b_n(B \times C) = \sum_{k+l=n} b_k(B)\, b_l(C)$$

where b_n are the Betti numbers.

If we form the Betti polynomials

$$b(B, z) = \sum_{n \geq 0} z^n b_n(B),$$

Corollary 7.10 implies that

$$b(B \times C, z) = b(B, z)\, b(C, z).$$

Consider now the following fiber space: take a topological space B, a point $b \in B$ and let X be the space of all curves in B starting at b, with the usual topology. Of course $p: X \to B$ assigns its end point to each curve.

Let $\phi : A \to X$, and $\psi = p\phi$. For a given homotopy ψ_t of ψ, put $\phi_t(a)$ = curve $\phi(a)$ followed by $\psi_t(a)$. This provides a lifting. So $X \xrightarrow{p} B$ has the lifting homotopy property. Furthermore, X has the homotopy type of a point (just shrink each curve to the point b). Therefore $H_n(X) = 0$ for $n > 0$.

Returning to Theorem 7.9, set $Z^{k,l} = H_k(B, H_l(C, G))$ for $k, l \geqq 0$, and denote by Z the whole double sequence $\{Z^{k,l}; k, l \geqq 0\}$. More generally, assume we have two arbitrary double sequences of Abelian groups, Z and $\hat{Z}$. Then we make the following

7.11. Definition: We say that $\hat{Z}$ is derived from Z by an r-boundary operation if there exists a "boundary" operator $d: \Sigma \oplus Z^{k,l} \to \Sigma \oplus Z^{k,l}$ such that $d^2 = 0$,

(*)
$$d: Z^{k,l} \to Z^{k-r,\, l+r-1}$$

and

$$\hat{Z}^{k,l} \cong \frac{\{dz = 0\} \cap Z^{k,l}}{dZ \cap Z^{k,l}}.$$

(It should be understood that $Z^{k,l}$ is the trivial group for k or $l < 0$.) d is called an operator of type r. In this case we shall write $\hat{Z} = \mathscr{H}_r(Z)$.

Observe that for r large and $k + l$ small, $\hat{Z}^{k,l} = Z^{k,l}$, because in this case $\{dz = 0\} = Z^{k,l}$ and $dZ = \{0\}$.

7.12. Lemma: If we have operators d_i of type $i = 2, 3, \ldots$ and starting with Z, sequences $\mathscr{H}_2(Z)$, $\mathscr{H}_3(\mathscr{H}_2(Z))$ of groups, etc., the limit

$$\mathscr{H}_\infty(Z) = \lim_n \mathscr{H}_n(\mathscr{H}_{n-1} \cdots (\mathscr{H}_2(Z)) \ldots)$$

exists.

This follows from the above observation.

Next we quote the following fundamental theorem on the homology of fiber space, but without giving its proof.

7.13. Theorem (Leray-Serre): [Cf. Serre, "Homologie Singulière des Espaces Fibrés", Ann. Math. 54 (1951).] *Let $X \xrightarrow{p} B$ be a fiber space, with B connected and simply connected, and connected fiber $F = p^{-1}(b)$. Put $Z^{k,l} = H_k(B, H_l(F))$. Then $H_n(X)$ has a composition series with factors $\hat{Z}^{k,l}$, $k + l = n$, such that $\hat{Z} = \mathscr{H}_\infty(Z)$.* (A composition series for G is a sequence of subgroups G_i of G such that $G = G_0 \supseteq G_1 \supseteq \cdots \supseteq 0$, and the factors are the groups G_i/G_{i+1}.)

Let us consider once more the fiber space $X \xrightarrow{p} B$ of curves beginning at $b \in B$, with fiber $F = p^{-1}(b) = \Omega(B)$. As we said before, all the homology groups of X are zero for $n > 0$. As in Theorem 7.13, put $Z^{k,l} = H_k(B, H_l(\Omega(B)))$, where $Z^{k,0} = H_k(B)$. Suppose that $H_n(B)$ is the first non-vanishing homology group of B of positive dimension (see diagram below). If $n > 2$, by Theorem 7.13, $Z^{0,1}$ must be zero, because this does not change when homology with respect to an $r/r \geqq 2$ boundary operation is taken; since the final result must be 0, all $\hat{Z}^{k,l}$ being zero, $H_1(\Omega)$ itself must vanish. This implies that all the $Z^{k,l}$ in the column of $H_1(\Omega)$ are 0. Similarly, if $n > 3$, all the groups in the column of $H_2(\Omega)$ are zero. Using these remarks, we may prove the following theorem.

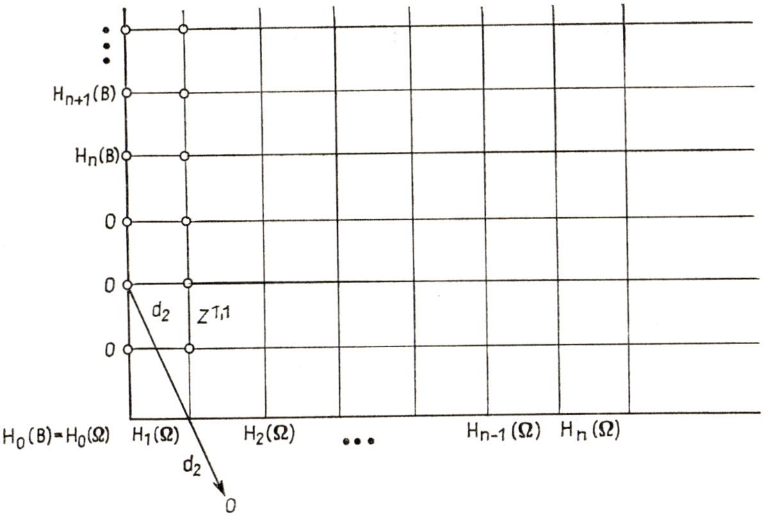

7.14. Theorem: *If B is connected and simply connected, the first non-vanishing homology group $H_n(B)$ of positive dimension is isomorphic to the first non-vanishing homology group of positive dimension of $\Omega(B)$, which is $H_{n-1}(\Omega(B))$.*

Thus

$$H_n(B) \simeq H_{n-1}(\Omega(B)).$$

Proof: Suppose, for example, that $n = 3$. After homology with respect to the 2-boundary operation is taken, $Z^{3,0}$ remains the same, for $Z^{1,1}$ is zero by the above remark. The same is true of $Z^{0,2}$. Taking homology with respect to the 3-boundary operation may change both groups, but all the other homologies leave invariant the groups in the places (3, 0) and (0, 2). But the limit groups $H_\infty(Z)^{3,0}$ and $H_\infty(Z)^{0,2}$ must be zero, so the 3-boundary homology gives us zero in both places. In other words, the sequence

$$0 \xrightarrow{d_3} H_3(B) \xrightarrow{d_3} H_2(\Omega) \xrightarrow{d_3} 0$$

is exact, which proves the theorem.

Q.E.D.

7.15. Corollary *(Hurewicz): If B is connected and simply connected, the first non-vanishing homology group of positive dimension, $H_n(B)$ is isomorphic to the first non-vanishing homotopy group of positive dimension $\pi_n(B)$.*

Proof:

$$H_n(B) \simeq H_{n-1}\,(\Omega(B) \simeq H_1\,(\Omega^{n-1}(B)) \simeq \pi_1\,(\Omega^{n-1}(B)) \simeq \pi_n(B).$$

Q.E.D.

Now assume that B is a finite dimensional space. Consider homology groups with real coefficients and let $D^{k,l} = \dim Z^{k,l} = b_k(B)\,b_l(\Omega(B))$, where the b_k are Betti numbers.

Suppose that $\Omega(B)$ has only finitely many non-vanishing Betti numbers; let $b_r(\Omega(B))$ be distinct from zero, and $b_l(\Omega(B)) = 0$ for $l > r$. Similarly, let $b_n(B) \neq 0$ and $b_k(B) = 0$ for $k > n$. Then $D^{r,n}$ is different from zero, and remains fixed throughout the sequence of homologies of Lemma 7.12 and Theorem 7.13 (same argument as before). But this is a contradiction, for the final result gives the trivial homology of the path-space X and hence must be 0. Thus $\Omega(B)$ always has infinitely many non-vanishing homology groups. Suppose next that one of the numbers, say, $b_s(\Omega(B))$, is infinite, and that for $l < s, b_l\,(\Omega(B))$ is finite. Then the number at the node $(s, 0)$ of the above diagram remains infinite throughout the sequence of homologies which is again a contradiction. We have thus proved

7.16. Theorem: *If B is connected, simply connected and finite dimensional, $\Omega(B)$ has infinitely many non-vanishing real homology groups and all of them have finite dimensions.*

Q.E.D.

The space $\Omega(B)$ is an example of the more general concept of a "group-like space".

7.17. Definition: *Let X be a topological space. Then X is called a group-like space if there is a binary operation defined on it, a distinguished element*

called the identity, and a mapping $x \to x^{-1}$ such that all the properties defining a group are satisfied up to homotopy (e.g. $m \to m \cdot e \sim$ identity).

For group-like spaces we have the following theorem of Hopf, which we quote from the ciled paper of Serre but shall not prove.

7.18. Theorem: *The cohomology ring with real coefficients of a group-like space with finite dimensional homology groups is the direct product of a polynomial algebra and an exterior algebra.*

7.19. Corollary: *Under the above hypotheses, if the group-like space X has infinitely many non-vanishing Betti numbers, the cup-length of X equals ∞.* (See § 2 of Chapter 5 for the definition of cup-length.)

7.20. Corollary: *If B is connected, simply connected and finite dimensional, then*

$$\text{cup-length } (\Omega(B)) = \infty .$$

[Compare with Theorem 5.16.]

7.21. Corollary: *Under the above hypotheses, any two points of a Riemannian manifold B are connected by indefinitely long geodesics.*

Remark: It can be proved that if B is compact, simple connectedness is not necessary.

C. The Homotopy of Some Lie Groups

We first recall the definition and some properties of the unitary group. For more details, see Milnor's book on Morse theory. The unitary group $U(n)$ is the group of all $n \times n$ complex matrices preserving the inner product in C^n, or equivalently, the group of all $n \times n$ complex matrices such that $UU^* = I$, where U^* is the conjugate transpose of U.

This is a Lie group, and the tangent space at the identity I is the space of matrices $\{iH\}$, where H is hermitian, i.e.: $H = H^*$. Analogously, the tangent space at U_0 is the space of matrices $\{iU_0H\} = \{iHU_0\}$. The matrix exponential function defined by

$$\exp A = I + A + \frac{A^2}{2!} + \frac{A^3}{3!} + \cdots$$

coincides with the exponential function defined on the Lie algebra $\{iH\}$ with values in the Lie group $U(n)$. The scalar product

$$(A, B) = Re \text{ trace } (AB^*)$$

defines a Riemannian structure on $U(n)$.

The geodesics beginning at I are the curves of the form $\sigma(t) = \exp(itH)$, with H hermitian. We say that $\sigma(t)$ *has H as initial velocity*.

Our aim now is to determine at which points of the tangent space at the origin, that is, at which hermitian matrices H, the exponential function has a vanishing Jacobian. The image of these points under the exponential is the set of conjugate points to the identity I.

In general, let f be any analytic function of a matrix. Then f has a Cauchy integral representation,

$$f(M) = \frac{1}{2\pi i} \oint \frac{f(z)}{z - M}\, dz.$$

If $\delta f(M, N)$ denotes the first variation of f at the point M applied to N, we have:

$$\delta f(M, N) = \frac{1}{2\pi i} \oint f(z)\, \delta\, [(z - M)^{-1}, N]\, dz. \tag{1}$$

On the other hand,

$$\delta\, (z - M)^{-1} = (z - M)^{-1}\, \delta M\, (z - M)^{-1}. \tag{2}$$

Now consider the operators $\varrho(A)$ and $\lambda(A)$, on matrices defined as right and left multiplication by A respectively. Since the mapping $A \to \varrho(A)$ is a homomorphism from the group of non-singular matrices to the group of non-singular linear operators on the linear space of all matrices, and similarly for λ, we obtain:

$$\varrho\, ((z - A)^{-1}) = (z - \varrho\, (A))^{-1}$$

and

$$\lambda\, ((z - A^{-1}) = (z - \lambda\, (A))^{-1}.$$

Hence from formula (2) we obtain

$$\delta\, ((z - M^{-1}) = (z - \varrho(M))^{-1}\, (z - \lambda(M))^{-1}\, \delta M,$$

and therefore using (1) it follows that

$$\delta f(M, N) = \frac{1}{2\pi i} \left\{ \oint f(z)\, (z - \varrho(M))^{-1}\, (z - \lambda(M))^{-1}\, dz \right\} (N). \tag{3}$$

Set

$$\phi\, (\xi_1, \xi_2) = \frac{1}{2\pi i} \oint f(z)\, \frac{1}{z - \xi_1}\, \frac{1}{z - \xi_2}\, dz.$$

Then

$$\delta f(M, N) = \phi\, (\varrho(M), \lambda(M))\, (N).$$

Moreover,

$$\frac{1}{z - \xi_1}\, \frac{1}{z - \xi_2} = \frac{1}{\xi_1 - \xi_2} \left(\frac{1}{z - \xi_1} - \frac{1}{z - \xi_2} \right),$$

so

$$\phi(\xi_1, \xi_2) = \frac{f(\xi_1) - f(\xi_2)}{\xi_1 - \xi_2}.$$

We now return to the function $f(H) = \exp(iH)$. In this case, we have

$$\delta \exp(iH) = \phi(\varrho(H), \lambda(H))\, \delta H,$$

where

$$\phi(\xi_1, \xi_2) = \frac{\exp(i\xi_1) - \exp(i\xi_2)}{\xi_1 - \xi_2} = \exp(i\xi_1)\frac{(1 - \exp(i(\xi_1 - \xi_2)))}{\xi_1 - \xi_2}.$$

But

$$\varrho(A) - \lambda(A) = Ad(A).$$

So finally we get the formula

$$\delta \exp(iH) = \exp(iH)\, \psi(Ad(H))\, \delta H,$$

where

$$\psi(z) = \frac{1 - \exp(iz)}{z}.$$

Furthermore, the eigenvalues of ψ (Ad (H)) are equal to ψ (eigenvalues of Ad (H)). The zeros of ψ are $z = 2\pi n$, $n = 0, \pm 1, \pm 2, \ldots$ Hence the matrices H which give rise to conjugate points to the identity, are those whose Ad (H) has an eigenvalue of the form $2\pi n$, $n = 0, \pm 1, \pm 2, \ldots$ Now, the eigenvalues of Ad (H) are differences of the eigenvalues of H, hence the matrices we are looking for are those having eigenvalues differing by $2\pi n$, $n = 0, \pm 1, \pm 2, \ldots$

All these calculations apply to the Special Unitary group $SU(n)$ also, but in this case the tangent space to the indentity, that is, the Lie algebra, is the space of matrices $\{iH\}$ where H is hermitian and has trace 0.

The following considerations apply to the group $U(2m)$ or the group $SU(2m)$.

We choose an element E near $-I$ having the form

$$E = \begin{pmatrix} \exp(i(\pi + \varepsilon_1)) & & & & 0 \\ & \exp(-i(\pi + \varepsilon_1)) & & & \\ & & \exp(i(\pi + \varepsilon_2)) & & \\ & & & \exp(-i(\pi + \varepsilon_2)) & \\ 0 & & & & \ddots \end{pmatrix}.$$

We wish to study the geodesics joining I and E. Therefore we have to find the solutions of $\exp(iH) = E$. But such an element H commutes with E, and since E is diagonal with distinct entries, H too must be diagonal. Set

$$H = \begin{pmatrix} h_1 & & & \\ & h_2 & & \\ & & \cdot & \\ & & & \cdot \\ & & & & \cdot \end{pmatrix}.$$

Then $\exp(ih_1) = \exp(i(\pi + \varepsilon_1))$; $\exp(ih_2) = \exp(i(\pi - \varepsilon_1))$; etc. Hence,

$$h_1 = \pi + \varepsilon_1 + 2\pi n_1 = \varepsilon_1 + (2n_1 + 1)\pi,$$

$$h_2 = \pi - \varepsilon_1 + 2\pi n_2 = -\varepsilon_1 + (2n_2 + 1)\pi, \quad \text{etc.}$$

This means that the h_i are of the form:

$$(2k + 1)\pi \pm \varepsilon.$$

The length of the geodesic with initial velocity H is

$$L = \left|\frac{d}{dt}\exp(itH)\right| = (tr\,(HH^*))^{1/2} = (tr(H^2))^{1/2}$$

$$= \{\Sigma\,((2n_j + 1)\pi \pm \varepsilon)^2\}^{1/2}.$$

Choosing ± 1 for the coefficients of π, we obtain 2^{2m} geodesics of minimal length $\sim \pi\sqrt{2m}$.

The next shortest geodesics are obtained when all coefficients but one are ± 1, and one of them is ± 3; then the length is $\sim \pi\sqrt{2m + 8}$.

Conjugate points to the identity appear along a geodesic when $t\,(h_i - h_j) = 2\pi n$ for some t, $0 < t < 1$, and for some $h_i \neq h_j$. Given h_i and h_j there are $\left[\frac{h_i - h_j}{2\pi}\right]$ ([ ] = integer part) conjugate points. The total number of conjugate points along the geodesic is therefore

$$\sum_{h_i \neq h_j}\left[\frac{h_i - h_j}{2\pi}\right] = 2\sum_{h_i > h_j}\left[\frac{h_i - h_j}{2\pi}\right].$$

But $h_j = \pi(2n_j + 1) \pm \varepsilon$. Hence the total number of conjugate points is $\geqq 2\sum_{n_j > n_j}(n_j - n_i - 1)$ for ε small enough.

Consider now the special case of the Special Unitary group $SU(2m)$. Then we have

7.22. Lemma: Unless m of the n_j's in the formula for the h_j's are 0 and the m others are -1, the geodesic with initial velocity H passes through at least $2(m+1)$ conjugate points to the identity.

Proof: Since the trace of H is zero, $\Sigma h_j = \pi(\Sigma 2n_j + 2m) = 0$, so $\Sigma n_j = -m$. Thus there are two possibilities: either $\sum_{n_j<0} n_j < -m$, or $\sum_{n_j<0} n_j = -m$. In the first case, there is at least one positive n_j, call it n_1, so that if N denotes the number of conjugate points

$$N \geqq 2 \sum_{n_j>n_i} (n_j - n_i - 1) \geqq 2 \sum_{n_j<0} (n_1 - n_j - 1) \geqq 2 \sum_{n_j<0} -n_j \geqq 2(m+1).$$

In the second case $\sum_{n_j<0} n_j = -m$ there are no positive n_j's. If our hypothesis is violated, some negative n_j must be less than -1, so the number of n_j's equal to zero is $\geqq m+1$. Now,

$$N \geqq 2 \sum_{\substack{n_k=0 \\ n_j<-1}} (n_k - n_j - 1) \geqq 2 \quad (\text{number of } n\text{'s equal to } 0) \geqq 2(m+1).$$

Q.E.D.

7.23. Corollary: All the geodesics joining I and E of non-minimal length contain at least $2(m+1)$ conjugate points and therefore have index $\geqq 2(m+1)$.

7.24. Corollary: For the loop space $H_1(SU(2m), I, E)$ and the length function J, the relative homotopy group

$$\pi_j\left(H_1, \left\{J \leqq \pi\sqrt{2m} + \delta\right\}\right) = 0$$

for all $j \leqq 2m+1$ and δ small enough.

The proof is an application of Corollary 7.4.

We consider now the geodesics joining I and $-I$ in $SU(2m)$. If H is the initial velocity of such a geodesic, H has eigenvalues $(2n_j+1)\pi$, and the length of the corresponding geodesic is

$$L = \left[\sum_{j=1}^{2m} ((2n_j + 1)\pi)^2\right]^{1/2}.$$

We obtain the geodesics of minimal length when all the coefficients $2n_j + 1$ of π, are ± 1. This length is $\sqrt{2m}\,\pi$, and the other geodesics have length $\geqq \pi\sqrt{2m+8}$. For the geodesics of minimal length, the fact that trace $(H) = 0$ implies that there are m eigenvalues equal to $+\pi$ and m eigenvalues equal to $-\pi$. In this case, the matrix H is completely determined by giving the subspace of eigenvectors of the eigenvalue π, the other subspace corresponding to $-\pi$ being orthogonal to the first one. Therefore we have a homeomorphism between the manifold of minimal geodesics joining I to $-I$ in $SU(2m)$ and the Grassmann manifold $G_m(2m)$ of m-dimensional linear subspaces of C^{2m}.

We shall now prove the following theorem due to Bott (compare to Lemma 22.5 of Milnor's *Morse Theory*):

7.25. Theorem (Bott): *Let M be a complete Hilbert manifold, f a smooth function satisfying the P-S condition. Suppose that the set $\{f \leqq a\}$ has only one critical level c, with critical set $K = f^{-1}(c)$, and assume that K is a finite dimensional submanifold of M. Then*

$$\pi_k(\{f \leqq a\}, K) = 0 \quad \text{for all } k,$$

provided that f is bounded below.

For the proof we need the following lemma.

7.26. Lemma: Under the hypotheses above, if $\{x_n\}$ is a sequence such that $f(x_n) \to c$, then there exists a subsequence $\{x_{n_i}\}$ such that $x_{n_i} \to x_\infty \in K$.

Proof: Take c to be 0. Consider the vector field $v = -\nabla f$, and let $\sigma(x, t)$ be the flow of v ($\sigma(x, 0) = x$; see Definition 4.44). For each n, let t_n be the first value of t for which

$$\|\nabla f(\sigma(x_n, t))\| \leqq \frac{1}{n}.$$

We prove the existence of such a t_n as follows.

If $g(t) = f(\sigma(t))$, $\sigma(t)$ being any solution of $\sigma'(t) = v(\sigma(t))$, then

$$g'(t) = df_{\sigma(t)}(\sigma'(t)) = df_{\sigma(t)}(-\nabla f_{\sigma(t)}) = -\|\nabla f_{\sigma(t)}\|^2,$$

so $g(t)$ is monotone decreasing. Since g is bounded below, its derivative cannot be bounded away from zero for positive t. Thus t_n exists. Set $y_n = \sigma(x_n, t_n)$. Then we have

$$f(y_n) \leqq f(x_n).$$

By the P-S condition, there is a subsequence $\{y_{n_j}\}$ which tends to a critical point y_∞. Since 0 is the only critical level, $f(y_{n_j}) \to f(y_\infty) = 0$. Assume without loss of generality that $nf(x_n) \to 0$. Then, $nf(y_n) \to 0$ also. But

$$d(x_{n_j}, y_{n_j}) = d(\sigma(x_{n_j}, 0), \sigma(x_{n_j}, t_{n_j})) \leqq \int_0^{t_{n_j}} \|\sigma'(x_{n_j}, t)\| \, dt$$
$$= \int_0^{t_{n_j}} \|\nabla f(\sigma(x_{n_j}, t))\| \, dt.$$

In the interval $(0, t_{n_j})$, $\|\nabla f(\sigma(x_{n_j}, t))\| > 1/n_j$, whence the last expression is less than

$$(*) \qquad n_j \int_0^{t_{n_j}} \|\nabla f(\sigma(x_{n_j}, t))\|^2 \, dt.$$

Now, $\frac{d}{dt} f(\sigma(t)) = -\|\nabla f(\sigma(t))\|^2$. So (*) equals

$$n_j \int_0^{t_{n_j}} -\frac{d}{dt} f(\sigma(x_{n_j}, t))\, dt = n_j \left(f(x_{n_j}) - f(y_{n_j})\right)$$

which tends to zero.

Thus $\{x_{n_j}\}$ also tends to $y_\infty \in K$.

Q.E.D.

Proof of Theorem 7.25: By our hypotheses, K has a neighborhood N in M homeomorphic to $K \times$ disc. [Cf. O. Hanner, "Some theorems on absolute neighborhood retracts", *Arkiv för Matematik*, Vol. 1, (1950), pp. 389–408.]

Assume that S is a topological space and $\phi: S \to \{f \leqq a\}$ is continuous. Using Morse Theory it follows that for any $\varepsilon > 0$, ϕ is homotopic to a ϕ_ε such that $c - \varepsilon \leqq f(\phi_\varepsilon(S)) \leqq c + \varepsilon$ [because $\{f \leqq a\}$ and $\{f \leqq c + \varepsilon\}$ have the same homotopy type]. Therefore there is a homotopic ϕ_1 such that $f(\phi_1(S)) \leqq c + \varepsilon$ and $\phi_1(S) \subseteq N$. If this were not the case, we would construct a sequence $\{x_n\}$ such that $f(x_n) \to c$ and $x_n \notin N$ for all n, contradicting Lemma 7.26. Since N is homeomorphic to $K \times$ disc it follows by squeezing the disc that ϕ_1 is homotopic to a function ϕ_2 with values in K.

Q.E.D.

We return to our study of the groups $U(2m)$ and $SU(2m)$. By Corollary 7.24, any map from a space X of dimension $< 2m + 1$ into $H_1(SU(2m), I, E)$ can be pushed down homotopically into a map of X with values in curves whose length is $\leqq \pi\sqrt{2m} + \varepsilon$, for any $\varepsilon > 0$. The same result is true if we consider instead the space of loops $H_1(SU(2m), I, -I)$; the points $-I$ and E being joined by an unique minimal geodesic one can prove that $H_1(SU(2m), I, E)$ and $H_1(SU(2m), I, -I)$ have the same homotopy type.

7.27. Lemma: For $k < 2m + 1$, $\pi_k(\Omega(SU(2m), I, -I), K) = 0$, where K is the manifold of minimal geodesics joining I and $-I$.

Proof: By the remark above, a map from the k-cube into $\Omega(SU(2m), I, -I)$ can be pushed down to a map into the space Ω_0 of curves of length at most

$\pi\sqrt{2m+\varepsilon}$. But by Bott's theorem, the space Ω_0 has vanishing homotopy groups relative to K.

Q.E.D.

7.28. Corollary: For $k < 2m+1$, $\pi_k(\Omega(SU(2m), I, -I)) \sim \pi_k(G_m(2m))$.

Proof: By Theorem 7.2, the sequence

$$\cdots \to \pi_k(X, A) \to \pi_{k-1}(A) \to \pi_{k-1}(X) \to \pi_{k-1}(X, A) \to \cdots$$

is exact. The first and the last written terms are zero by Lemma 7.27, whence the middle terms are isomorphic, i.e.

$$\pi_k(\Omega(SU(2m), I, -I)) \simeq \pi_k(K).$$

But, as noted preceding Theorem 7.25, K is homeomorphic to the Grassmann manifold $G_m(2m)$.

Q.E.D.

7.29. Corollary (Bott's isomorphisms):

$$\pi_{k+1}(SU(2m)) \simeq \pi_k(G_m(2m)) \quad \text{for} \quad k \leqq 2m.$$

We now proceed to obtain corresponding results for the group $U(n)$. Let $X \xrightarrow{p} B$ be a fiber space, X and B connected.

Let $F = p^{-1}(b)$ be the fiber. Then p induces an isomorphism $p^*: \pi_k(X, F) \to \pi_k(B)$ [cf. Steenrod, *The Topology of Fiber Bundles*, or Hilton and Wylie, *Homology Theory*, pp. 288–289]. Using the exact sequence for homotopy we see that

$$\cdots \to \pi_k(B) \to \pi_{k-1}(F) \to \pi_{k-1}(X) \to \pi_{k-1}(B) \to \cdots$$

is exact [this is the so called exact bundle sequence].

If G is a connected Lie group, and H a subgroup of G, then G is a fiber bundle with base space G/H (the factor space of G by H). The projection p is the natural one. [This uses the existence of local cross sections of G over G/H. See Steenrod's book.] Consider now the inclusion $SU(n) \subset U(n)$. The factor space $U(n)/SU(n)$ is the unit circle C. Then:

$$\pi_k(U(n)) \simeq \pi_k(SU(n)), \; k > 2.$$

Next, consider the inclusion $U(n) \subset U(n+1)$. The factor space $U(n+1)/U(n)$ is the sphere S^{2n+1} [cf. Steenrod's book]. Hence,

$$\pi_k(S^{2n+1}) \to \pi_{k-1}(U(n)) \to \pi_{k-1}(U(n+1)) \to \pi_{k-1}(S^{2n+1})$$

is exact. So, for $k < 2n + 2$ we get:

$$\pi_k(U(n)) \simeq \pi_k(U(n+1))$$

(Stable homotopy groups). It is natural to write $\pi_k(U(\infty)) = \lim_{n\to\infty} \pi_k(U(n))$.

The space $U(m) \times U(m)$ is included in $U(2m)$, since the matrix

$$\left(\begin{array}{c|c} A(m) & 0 \\ \hline 0 & A(m) \end{array}\right)$$

is in $U(2m)$ for any $A(m) \in U(m)$.

It is easy to verify that the factor space $U(2m)/U(m) \times U(m)$ is homeomorphic to the Grassmann manifold $G_m(2m)$. Therefore

$$\pi_k(G_m(2m)) \to \pi_{k-1}(U(m) \times U(m)) \to \pi_{k-1}(U(2m)) \to \pi_{k-1}(G_m(2m))$$

is exact. This geometric fact justifies the following

Remark: Given a Lie group G and subgroups $H_1 \subset H_2$, G/H_1 has a bundle structure over G/H_2, with natural projection and fiber H_2/H_1. [Same proof as in the somewhat less general case considered before.] If we then consider the fiber space

$$U(2m)/U(m) \to U(2m)/U(m) \times U(m),$$

and use the exact bundle sequence we find that

$$\pi_k(U(2m)/U(m)) \to \pi_k(G_m(2m)) \to \pi_{k-1}(U(m)) \to \pi_{k-1}(U(2m)/U(m))$$

is exact.

7.30. Theorem: (Bott's Periodicity Theorem):

$$\pi_{k-1}(U(\infty)) \simeq \pi_{k+1}(U(\infty)) \quad \text{for } k \geqq 1.$$

Proof: First we prove that in the exact sequence noted above the first and the last groups are zero provided m is large enough. Indeed, we have seen already that the space $U(2m)/U(2m-1)$ is the sphere S^{4m-1}. Thus

$$\pi_k(U(2m)/U(2m-1)) = 0 \quad \text{for } m \text{ large.}$$

Similarly

$$\pi_k(U(2m-1)/U(2m-2)) = 0 \quad \text{for } m \text{ large, etc.,}$$

and we get

$$\pi_k(U(2m)/U(m)) = 0 \quad \text{for } m \text{ large,}$$

by the above remark.

Hence, by the exactness of the above sequence of homotopy groups, we get

$$\pi_k(G_m(2m)) \simeq \pi_{k-1}(U(m)) \quad \text{for } m \text{ large.}$$

Now,

$$\pi_{k-1}(U(m)) \simeq \pi_{k-1}(U(2m)) \quad \text{by stability, for } m \text{ large.}$$

Therefore

$$\pi_k(G_m(2m)) \simeq \pi_{k-1}(U(2m)) \quad \text{for } m \text{ large.}$$

Using Corollary 7.29 (Bott's isomorphisms), we may assert that

$$\pi_{k+1}(SU(2m)) \simeq \pi_k(G_m(2m)) \quad \text{for } m \text{ large.}$$

But we have seen that

$$\pi_k(U(n)) \simeq \pi_k(SU(n)) \quad \text{for } \; k > 2.$$

Thus

$$\pi_{k+1}(U(2m)) \simeq \pi_k(G_m(2m)) \quad \text{for } m \text{ large,}$$

and for $k > 1$. Hence, finally,

$$\pi_{k+1}(U(2m)) \simeq \pi_{k-1}(U(2m)) \quad \text{if } m \text{ is large.}$$

Q.E.D.

7.31. Corollary: The homotopy groups $\pi_k(U(\infty))$ are zero if k is even, and isomorphic to Z if k is odd.

Proof: Observe that

$$\pi_2(U(\infty)) \simeq \pi_2(SU(2)) = 0$$

$$\pi_3(U(\infty)) \simeq \pi_3(SU(2)) \simeq Z,$$

as $SU(2)$ is nothing but S^3.

Q.E.D.

CHAPTER VIII

Closed Geodesics on Compact Riemannian Manifolds

(Chapter by Hermann Karcher)

In this chapter closed geodesics on a Riemannian manifold M will be studied using infinite dimensional Morse Theory in much the same way as in Chapter IV where geodesics joining two fixed points were treated. We shall study a Hilbert manifold $H_1(S_1, M)$ of closed, sufficiently regular curves (H_1-curves) on M. The coordinate spaces for $H_1(S_1, M)$ are (as in Chapter IV) Hilbert spaces whose elements are H_1-vector fields along curves on M. In Chapter IV one defined scalar products for the coordinate spaces (following Palais) with the aid of Nash's embedding theorem. In this chapter (cf. Theorem 8.6) we use instead Klingenberg's intrinsic scalar product (first introduced in a lecture given in Bonn) which depends only on the Riemannian structure of M. In Theorem 8.9 we prove that this scalar product and various other possible scalar products on the Hilbert spaces of H_1-vector fields lead to equivalent norms. The differentiable structure of $H_1(S_1, M)$ and useful coordinate systems are discussed in 8.11 to 8.18. Theorem 8.19 states the differentiability of the energy function and in 8.20 we introduce a Riemannian metric for $H_1(S_1, M)$ based on Klingenberg's intrinsic scalar products. These developments are somewhat more complicated than the corresponding ones in Chapter IV since it does not seem possible to obtain the intrinsic Riemannian metric of $H_1(S_1, M)$ via an embedding $M \subseteq R^N$. In 8.22 to 8.26 we carry out an auxiliary discussion of differentiable curves on $H_1(S_1, M)$ and their representation on M. In the second half of the present chapter our use of the intrinsic scalar product allows simpler proofs than in Chapter IV; it also seems that notions such as the gradient vector field of J on $H_1(S_1, M)$ can be more readily interpreted in terms of M. We prove a few geometric results concerning the Riemannian structure of $H_1(S_1, M)$ in 8.27 to 8.33.

Lemma 8.34 contains basic estimates which we need to derive the explicit formula for the first derivative of the energy J in 8.35 and to prove the validity of the Palais-Smale condition for J in 8.41. In 8.36 to 8.39 we introduce the gradient of J as a vector field on $H_1(S_1, M)$ and identify the critical points of J as the closed geodesics on M. In 8.41 to 8.50 standard arguments from infinite dimensional Morse Theory are used to show the existence of at least one nontrivial closed geodesic on every compact C^6-Riemannian manifold. In 8.48 we prove that those flow lines of the gradient deformation (cf. 8.43), which start at points f with sufficiently small energy $J(f) < e$ have uniformly bounded length. As an immediate consequence we obtain the important result 8.47 that $J^{-1}(0)$ is deformation retract of $J^{-1}([0, e])$.

We conclude with a summary of recent developments.

Preliminaries. M will be a compact Riemannian manifold of class C^k ($k \geqq 6$). (Metric completeness rather than compactness is sufficient for most of our general developments but not for the desired application to closed geodesics.) M_p denotes the tangent space to M at p, TM the tangent bundle of M. The scalar product in M_p is denoted by $g(p)(v, w)$ or more briefly by (v, w); in local coordinates on M the metric is written $g_{ik}(p)\, v^i w^k$. The distance on M induced by this infinitesimal metric (cf. Chapter VI) is called $d_M(p, q)$.

Absolutely continuous curves (resp. vector fields) with locally square integrable derivatives will be called H_1-curves (resp. H_1-vector fields). For H_1-curves we may define an energy integral J and a length L_M as follows.

$$J(f) = \tfrac{1}{2} \int \big(\dot{f}(t), \dot{f}(t)\big)\, dt, \quad L_M(f) = \int \big(\dot{f}(t), \dot{f}(t)\big)^{1/2}\, dt.$$

We shall be interested in closed H_1-curves parametrized by the interval $[0, 1]$ (not necessarily proportionally to arc length). Hence we find it useful to define the following space:

$$H_1(S_1, M) = \{f \mid f\colon [0, 1]/\{0, 1\} \to M \quad \text{and} \quad J(f) < \infty\}.$$

(We always identify the circle S_1 with the factor space $[0, 1]/\{0, 1\}$.)

The covariant derivative of a vector field $v(t)$ along $f(t)$ will be written Dv/dt; this derivative is given in local coordinates on M by the formula

$$\frac{Dv^i}{dt} = \frac{dv^i}{dt} + \Gamma^i_{jk}(f(t)) \cdot \dot{f}^j(t)\, v^k(t)$$

(where $\Gamma^i_{jk} \in C^{k-2}$ are the Christoffel symbols). Differential equations with square-integrable coefficients can be treated by the Picard-Lindelöf iteration scheme. Using this fact and the last formula we see that Levi-Civita parallel transport is well defined along any curve $f \in H_1(S_1, M)$. In particular a parallel vector field along such a curve is absolutely continuous, and for any continuous vector field along f the following statement holds: dv/dt is locally square integrable if and only if Dv/dt is locally square integrable.

The *exponential map* available on the manifold M may be described as follows. For $0 \neq v \in M_p$, let $c\colon [0, \infty) \to M$ be the geodesic ray starting at p with tangent vector v, such that its parameter t is proportional to arc length s and $ds/dt = |v|_{M_p}$. Define $\exp(v) = c(1)$. Then $\exp\colon TM \to M$. We denote the restriction $\exp|_{M_p} = \exp_p$. If convenient, we write $\exp_p = \exp$. It follows immediately from the differential equation for geodesics (i.e., from $D/\partial t\, \dot{c} = 0$) that $\exp_p(t \cdot v) = c(t)$, in other words that $\exp_p$ is radial isometric. Since $\Gamma^l_{mn} \in C^{k-2}$, it follows that $\exp\colon TM \to M$ is a C^{k-2} map, and the differential equation for geodesics also shows that the linear map induced by $\exp_p$ at the origin of M_p is the identity map.

Geodesic parallel coordinates on M are easily defined in terms of the exponential map exp. Given a geodesic arc $c\colon [0, T] \to M$ (arc length t as parameter), choose an orthonormal n-frame $F_0 = \{c(0), v_2(0), \ldots, v_n(0)\}$ in $M_{c(0)}$ and define n-frames F_t in $M_{c(t)}$ by parallel transport of F_0 along c. The map $[0, T] \times R^{n-1} \to M$ given by $(t, u^2, \ldots, u^n) \to \exp_{c(t)}\left(\sum_{i=2}^{n} u^i \cdot v_i(t)\right)$ is C^{k-2} and the induced linear map at $(t, 0, \ldots, 0)$ is the identity. Therefore a neighborhood of $[0, T] \times \{0\}$ is mapped C^{k-2} diffeomorphically onto a tubular neighborhood of c in M. The inverse map gives the desired coordinates. We have $g_{ik}(c(t)) = \delta_{ik}$ and $\Gamma^i_{jk}(c(t)) = 0$. Since M is compact these remarks prove the following lemma.

8.1. Lemma: There exists $\varepsilon_p > 0$ such that the geodesic parallel coordinates just defined are valid in the ε_p-tubular neighborhood of any geodesic arc which is sufficiently short so that its ε_p-tube does not cover any point twice. Moreover there exist constants $C > 0$ and $0 < m_1 \leqq 1 \leqq m_2 < \infty$ such that for parallel coordinates in any ε_p-tube we have

$$(1) \qquad |\Gamma^i_{jk}| \leqq C \quad \text{and} \quad m_1 \sum_{i=1}^{n} (v^i)^2 \leqq g_{ik} v^i v^k \leqq m_2 \sum_{i=1}^{n} (v^i)^2.$$

A consequence of the second inequality in (1) is given in

8.2. Lemma: If (,) and ((,)) are two scalar products such that $m_1((v, v)) \leqq (v, v) \leqq m_2((v, v))$ for all v and if $m_1 \leqq 1 \leqq m_2$ then

$$|(v, w) - ((v, w))| \leqq 16\,(m_2 - m_1)\sqrt{((v, v)) \cdot ((w, w))}.$$

Proof: Using $(v, w) = \frac{1}{4}((v + w, v + w) - (v - w, v - w))$ one first gets

$$|(v, w) - ((v, w))| \leqq 8\,(m_2 - m_1)\,(((v, v)) + ((w, w))).$$

Now $(v, w) - ((v, w)) = b\,(v, w)$ is bilinear and

$$|b\,(v, w)| \leqq c\,(|v|^2 + |w|^2)$$

implies

$$|b\,(v, w)| \leqq 2\,|v| \cdot |w|,$$

since for $\lambda \neq 0$ we have

$$|b\,(v, w)| = \left|b\left(\lambda v, \frac{1}{\lambda}\,w\right)\right| \leqq c\left(\lambda^2\,|v|^2 + \frac{1}{\lambda^2}\,|w|^2\right).$$

Q.E.D.

The following result is well known for H_1-curves in R^N.

8.3. Lemma: Every subset of $H_1(S_1, M)$ on which the energy integral is bounded consists of equicontinuous curves on M.

Proof:

$$d_M(f(t), f(t_0)) \leqq L_M(f)|_{t_0}^t = \int_{t_0}^t (\dot{f}(\tau), \dot{f}(\tau))^{1/2}\,d\tau$$

$$\leqq \sqrt{|t - t_0|} \cdot \left(\int_{t_0}^t (\dot{f}(\tau), \dot{f}(\tau))\,d\tau\right)^{1/2} \quad \text{(by the Cauchy-Schwarz inequality)}$$

$$\leqq \sqrt{|t - t_0|} \cdot \sqrt{2J(f)}\,.$$

8.4. Corollary: $L_M^2(f) \leqq 2J(f)$, and $L_M^2(f) = J(f)$ if and only if f is parametrized proportional to arc length, i.e., $(\dot{f}(\tau), \dot{f}(\tau)) = \text{const.}$

We next wish to define a differentiable structure on $H_1(S_1, M)$ with the aid of coordinate spaces which have a geometric interpretation on M.

8.5. Definition: For any $f \in H_1(S_1, M)$ consider the set of H_1-vector fields along f:

(2) $$H_1(S_1, TM_f) = \{v \mid v : [0, 1]/\{0, 1\}\} \to TM$$

such that $v(t) \in M_{f(t)}$ and v is absolutely continuous and has locally square integrable derivatives.

8.6. Theorem: $H_1(S_1, TM_f)$ is a separable Hilbert space with the scalar product $\langle v, w \rangle$ defined below.

Proof: $H_1(S_1, TM_f)$ is obviously a vector space. For $v, w \in H_1(S_1, TM_f)$ define

$$\langle v, w\rangle = \int_0^1 \left\{(v(t), w(t)) + \left(\frac{Dv}{dt}, \frac{Dw}{dt}\right)\right\} dt. \tag{3}$$

(Of course, D/dt denotes the covariant derivative along f.) It is clear from (2) of Definition 8.5 that we have $\langle v, v\rangle < \infty$, $\langle w, w\rangle < \infty$. Moreover using $|(v(t), w(t))| \leqq |v(t)| \cdot |w(t)|$ (in $M_{f(t)}$) and the Cauchy-Schwarz inequality, it follows that

$$\begin{aligned}\langle v, w\rangle^2 &\leqq \left[\int_0^1 \left\{|v(t)| \cdot |w(t)| + \left|\frac{Dv}{dt}\right| \cdot \left|\frac{Dw}{dt}\right|\right\} dt\right]^2 \\ &\leqq \int_0^1 \left\{|v(t)|^2 + \left|\frac{Dv}{dt}\right|^2\right\} dt \cdot \int_0^1 \left\{|w(t)|^2 + \left|\frac{Dw}{dt}\right|^2\right\} dt \\ &= \langle v, v\rangle \cdot \langle w, w\rangle.\end{aligned}$$

Hence $\langle v, w\rangle$ is defined for all $v, w \in H_1(S_1, TM_f)$. Clearly $\langle v, w\rangle$ is bilinear and positive definite, and therefore a scalar product.

For the proof of completeness of the space $H_1(S_1, TM_f)$ we need the following definition and lemma.

8.7. Definition: For $v \in H_1(S_1, TM_f)$ put

$$\|v\|_\infty = \max_{t \in [0,1]} (v(t), v(t))^{1/2}.$$

8.8. Lemma: $\|v\|_\infty^2 \leq 2\, \langle v, v\rangle$.

Proof: The formula

$$\frac{d}{dt}(g(f(t))(v(t), w(t))) = \left(\frac{Dv}{dt}, w(t)\right) + \left(v(t), \frac{Dw}{dt}\right) \tag{4}$$

is well known for $f, v, w \in C^1$ and generalizes by an easy limit argument to all $f \in H_1(S_1, M)$ and $v, w \in H_1(S_1, TM_f)$. Now choose t_m such that $\|v\|_\infty^2 = (v(t_m), v(t_m))$ and note that

$$\begin{aligned}\|v\|_\infty^2 &= (v(t), v(t)) + \int_t^{t_m} \frac{d}{d\tau}(v(\tau), v(\tau))\, d\tau \\ &\leqq (v(t), v(t)) + 2\int_0^1 |v(\tau)| \cdot \left|\frac{Dv}{d\tau}\right| d\tau.\end{aligned}$$

Since the left side of this formula is independent of t and since $2a \cdot b \leqq a^2 + b^2$ we get

$$\|v\|_\infty^2 \leqq \int_0^1 (v(t), v(t))\, dt + \int_0^1 \left\{(v(\tau), v(\tau)) + \left(\frac{Dv}{d\tau}, \frac{Dv}{d\tau}\right)\right\} d\tau \leqq 2\langle v, v\rangle.$$

Q.E.D.

We may now complete the proof of Theorem 8.6 in regard to completeness and separability.

Since f is absolutely continuous we can subdivide f into finitely many subarcs f_ν such that each of them is representable in some geodesic parallel coordinates by functions $f_\nu^i(t)$ ($i = 1, \ldots, n = \dim M$ and $t \in I_\nu$) with $|f_\nu^i(t)| < \varepsilon_p$ (cf. Lemma 8.1) and $\cup I_\nu = [0, 1]$, $\sum |I_\nu| = 1$. If and only if $f \in H_1(S_1, M)$ will all the f_ν^i be H_1 functions. Since $J(f) < \infty$ we can also assume that these subarcs are so short that

$$\int_{t \in I_\nu} (\dot{f}(t), \dot{f}(t))\, dt < \frac{m_1^2}{8C^2 m_2 n^3},$$

where C, m_1, m_2 are the constants of 8.1 (1), and n is the dimension of M. We now consider the restriction of any $v \in H_1(S_1, TM_f)$ to I_ν, i.e. to a vector field along f_ν. The coordinates of v will be called v^i. Then by 8.1

$$\langle v, v\rangle\Big|_{I_\nu} = \int_{I_\nu} \left\{(v(t), v(t)) + \left(\frac{Dv}{dt}, \frac{Dv}{dt}\right)\right\} dt$$

$$\geqq m_1 \int_{I_\nu} \left\{\sum_{i=1}^n ((v^i(t))^2 + (\dot{v}^i(t) + \Gamma_{jk}^i(f_\nu(t))\, \dot{f}_\nu^j(t)\, v^k(t))^2)\right\} dt.$$

We use $(A + B)^2 \geqq \frac{1}{2}A^2 - B^2$ to obtain

$$\langle v, v\rangle\Big|_{I_\nu} \geqq m_1 \int_{I_\nu} \left\{\sum_i ((v^i)^2 + \tfrac{1}{2}(\dot{v}^i)^2 - (\Gamma_{jk}^i \dot{f}_\nu^j v^k)^2)\right\} dt$$

$$\geqq m_1 \int_{I_\nu} \left\{\sum_i \left((v^i)^2 + \tfrac{1}{2}(\dot{v}^i)^2\right) - C^2 n \left(\sum_{j,k} \dot{f}_\nu^j v^k\right)^2\right\} dt.$$

Note that by 8.1 (1) and 8.8

$$(\textstyle\sum v^k)^2 \leqq n \sum (v^k)^2 \leqq \dfrac{n}{m_1}(v(t), v(t)) \leqq \dfrac{2n}{m_1}\langle v, v\rangle\Big|_{I_\nu},$$

and note also that $\displaystyle\int_{I_\nu} \left(\sum \dot{f}_\nu^j\right)^2 dt \leqq \frac{n}{m_1}\int_{I_\nu} (\dot{f}(t), \dot{f}(t))\, dt$, so that

$$\langle v, v\rangle\Big|_{I_\nu} \geqq \frac{m_1}{2}\int_{I_\nu} \sum_i ((v^i)^2 + (\dot{v}^i)^2)\, dt - \frac{2C^2 n^3}{m_1}\langle v, v\rangle\Big|_{I_\nu}\int_{I_\nu} (\dot{f}(t), \dot{f}(t))\, dt.$$

By the choice of the I_ν we have

$$1 + \frac{2C^2n^3}{m_1}\int_{I_\nu} (\dot{f}(t), \dot{f}(t))\, dt < 1 + \frac{1}{4}$$

which implies

$$\langle v, v\rangle\Big|_{I_\nu} \geqq \frac{m_1}{3}\int_{I_\nu} \sum ((v^i(t))^2 + (\dot{v}^i(t))^2)\, dt.$$

Similarly, we may deduce

$$\langle v, v\rangle\Big|_{I_\nu} \leqq 2m_2\int_{I_\nu} \sum_i ((v^i)^2 + (\dot{v}^i)^2)\, dt + \frac{4C^2n^3m_2}{m_1^2}\langle v, v\rangle\Big|_{I_\nu}\cdot\int_{I_\nu} (\dot{f}, \dot{f})\, dt,$$

so that we obtain

$$\langle v, v\rangle\Big|_{I_\nu} \leqq 4m_2\int_{I_\nu} \sum ((v^i(t))^2 + (\dot{v}^i(t))^2)\, dt.$$

Therefore $\langle v, v\rangle\Big|_{I_\nu}$ and the norms $\int_{I_\nu} \sum ((v^i)^2 + (\dot{v}^i)^2)\, dt$ are equivalent. But the completeness and separability of H_1 in this latter norm is well-known. Q.E.D.

The Definition 8.5 of the vector space $H_1(S_1, TM_f)$ involves only the differentiable structure of M. In Theorem 8.6 we prove completeness with respect to a scalar product which is defined intrinsically in terms of a Riemannian metric on M (justifying the name intrinsic scalar product). It will be helpful in what follows to know that various possible scalar products on $H_1(S_1, TM_f)$, are in fact equivalent.

8.9. Theorem: (i) Let $g^{(1)}$ and $g^{(2)}$ be Riemannian metrics on M. The corresponding scalar products on the various spaces $H_1(S_1, TM_f)$ as defined in 8.6 (3) are then equivalent, i.e. $c_1\|v\|^{(1)} \leqq \|v\|^{(2)} \leqq c_2\|v\|^{(1)}$ for each $v \in H_1(S_1, TM_f)$ with constants depending only on the energy $J(f)$ of the curve f.

(ii) Let $M \subset R^N$ be a differentiable submanifold, so that $H_1(S_1, M)$ may be regarded as a closed submanifold of the Hilbert space $H_1(S_1, R^N)$ (Theorem 6.11), and so that the usual H_1-norm for $H_1(S_1, R^N)$ induces a scalar product on each tangent space $H_1(S_1, TM_f)$ of $H_1(S_1, M)$. Then any two embeddings induce scalar products in $H_1(S_1, TM_f)$ which define norms for these spaces which are uniformly equivalent on any subset of $H_1(S_1, M)$ of bounded energy $J(f)$.

(iii) Any two scalar products in $H_1(S_1, TM_f)$ of the sort described in (i) and (ii) above are uniformly equivalent on any subset of $H_1(S_1, M)$ of bounded energy.

Proof: We first prove (i). By compactness of M there exist constants $0 < m_1 < m_2 < \infty$ such that

$$m_1 g_{ik}^{(1)} v^i v^k \leqq g_{ik}^{(2)} v^i v^k \leqq m_2 g_{ik}^{(1)} v^i v^k.$$

This implies $m_1 J^{(1)}(f) \leqq J^{(2)}(f) \leqq m_2 J^{(1)}(f)$. This observation allows us to repeat the equivalence-of-norms proof given above as the second part of the proof of Theorem 8.6 with only minor changes. The number of subintervals I_ν which are needed in that proof can trivially be estimated in terms of a bound for $J(f)$; the other constants appearing in the proof of Theorem 8.6 are independent of f. All additional details are left to the reader.

We now prove (iii). Given an embedding $M \subseteq R^N$, consider the norm $\| \ \|^*$ on $H_1(S_1, TM_f)$ which is induced by the corresponding embedding $H_1(S_1, M) \subset H_1(S_1, R^N)$ as described in (ii). Note also that the Riemannian metric on M induced by the embedding $M \subset R^N$ defines an intrinsic norm $\| \ \|'$ on $H_1(S_1, TM_f)$. We claim that $\| \ \|'$ and $\| \ \|^*$ are equivalent, uniformly for any subset of $H_1(S, M)$ of bounded energy. Indeed, for $v \in H_1(S_1, TM_f)$ we have $(\|v\|')^2 = \int_0^1 (v(t), v(t))_{M_{f(t)}}\, dt + \int_0^1 \left(\frac{Dv}{dt}, \frac{Dv}{dt}\right)_{M_{f(t)}} dt$ while

$$(\|v\|^*)^2 = \int_0^1 (v(t), v(t))_{R^N}\, dt + \int_0^1 \left(\frac{dv}{dt}, \frac{dv}{dt}\right)_{R^N} dt.$$

Of course $(v(t), v(t))_{M_{f(t)}} = (v(t), v(t))_{R^N}$ by the choice of the metric for M.

Let the map $\Phi: M \to R^N$ define the embedding with which we are concerned. The *second fundamental* form of $M \subset R^N$ in terms of local coordinates for M is a symmetric matrix of vector valued functions of these coordinates, equal specifically to the projection of $\partial^2\Phi/\partial u^i \partial u^j$ onto the hyperplane of directions normal to M. We denote this form by $\vec{l}_{ij}$. (Cf. Milnor, Morse Theory.) The definition of the covariant derivative of a vector field v along a curve f (in local coordinates v^i, f^i) can be seen to imply the formula

$$\left(\frac{dv}{dt}, \frac{dv}{dt}\right)_{R^N} = \left(\frac{Dv}{dt}, \frac{Dv}{dt}\right)_{M_{f(t)}} + (\vec{l}_{ij} v^i \dot{f}^j, \vec{l}_{ij} v^i \dot{f}^j)_{R^N},$$

from which it is plain that $\| \ \|' \leqq \| \ \|^*$. On the other hand, letting K_m be the least upper bound for the principal curvatures of M in all directions, it follows by the compactness of M (the embedding being fixed) that $K_m < \infty$. Moreover, by the definition of the principal curvatures,

$$(\vec{l}_{ij}v^i\dot{f}^j, \vec{l}_{rs}v^r\dot{f}^s)_{RN} \leqq K_m^2 \, |v(t)|^2 \, |\dot{f}|^2.$$

Hence, using 8.8

$$(\|v\|^*)^2 \leqq (\|v\|')^2 + K_m^2 \, 2J(f) \, 2(\|v\|')^2.$$

This proves the equivalence of the norms $\| \ \|^*$ and $\| \ \|'$. Combining (i) and (iii), (ii) follows trivially; details are left to the reader. Q.E.D.

Remark: It is no coincidence that the constants c_1, c_2 such that $c_1 \|v\|^{(1)} \leqq \|v\|^{(2)} \leqq c_2 \|v\|^{(1)}$ in the preceding theorem depend on an upper bound for the energy $J(f)$. One can easily check by explicit calculations (carried out most simply on the sphere or on manifolds with vanishing curvature) that the best possible constants c_1 and c_2 may indeed approach 0 and ∞ respectively as $J(f) \to \infty$.

Our next developments will depend on Palais' Theorem 6.12, which we restate as follows.

8.10. Theorem: Let M and N be closed C^k-submanifolds of R^m and R^n respectively. Then $H_1(S_1, M)$ and $H_1(S_1, N)$ are closed C^{k-4}-submanifolds of $H_1(S_1, R^m)$ and $H_1(S_1, R^n)$ respectively (cf. Theorem 6.11). Let $\Phi: M \to N$ be a C^{k-2} map.

(i) It is an elementary result of differential topology that Φ can be extended to a C^{k-2} map from R^m to R^n. Then, by Theorem 6.8, the extended Φ gives rise to a C^{k-4} map of $H_1(S_1, R^m)$ into $H_1(S_1, R^n)$.

(ii) Consequently the map $\overline{\Phi}: H_1(S_1, M) \to H_1(S_1, N)$ defined by $\overline{\Phi}(f) = \Phi \circ f$ is a C^{k-4} map. Moreover $d\overline{\Phi}_f(v)(t) = d\Phi_{f(t)}(v(t))$ for $v \in H_1(S_1, TM_f)$.

8.11. Definition: As in Chapter VI we take as the differentiable structure of $H_1(S_1, M)$ that structure which it inherits as a submanifold of $H_1(S_1, R^N)$ in virtue of an embedding of $M \subset R^N$. It is clear from Theorem 8.10 (cf. also Theorem 6.11) that the differentiable structure thus defined does not depend on the embedding. Although the differentiable structure of $H_1(S_1, M)$ is independent of the Riemannian metric of M it will be of considerable advantage to have coordinate neighborhoods and maps on $H_1(S_1, M)$ available which are closely related to the Riemannian structure of M. We introduce such coordinates in the following definition and lemmas.

8.12. Lemma: The set $O(f) = \{v \mid v \in H_1(S_1, TM_f), \ \|v\|_\infty < \varepsilon\}$, is an *open* subset of the Hilbert space $H_1(S_1, TM_f)$.

Proof. Let $v \in O(f)$, i.e. let $\|v\|_\infty < \varepsilon$ so that $\|v\|_\infty \leqq \varepsilon - 2\delta$ for some $\delta > 0$. Then if $\langle w, w\rangle^{1/2} = \|w\| < \delta$ we have $\|w\|_\infty < 2\delta$ (cf. Lemma 8.8 above). Hence $\|v + w\|_\infty < \varepsilon$, so that the δ-ball around v is in $O(f)$. Q.E.D.

8.13. Definition: For $f, h \in H_1(S_1, M)$ put

$$d_\infty(f, h) = \max_{t \in S_1} d_M(f(t), h(t)); \tag{1}$$

given $f \in H_1(S_1, M)$ and $\varepsilon < \frac{1}{2}\varepsilon_p$ (cf. Lemma 8.1). Put

$$U(f) = \{h \mid h \in H_1(S_1, M) \quad \text{and} \quad d_\infty(f, h) < \varepsilon\}. \tag{2}$$

This set is introduced as a *standard coordinate neighborhood* of f, a definition justified by the two following lemmas. Note that $U(f)$ is an open subset of $H_1(S_1, M)$ since the original Riemannian metric of M and the metric induced by an embedding $M \subset R^N$ (as used in Definition 8.11) are equivalent and $V(f) = \{h \in H_1(S_1, R^N);\ \max_t d_{R^N}(h(t), f(t)) < \delta\}$ is open in $H_1(S_1, R^N)$ by the proof of Lemma 8.12.

8.14. Lemma: The following formula defines a 1–1 correspondence η between $U(f)$ and $O(f)$:

$$h(t) = \exp_{f(t)}(v(t)).$$

Proof: We have $|v(t)| = d_M(f(t), h(t))$ by the radial isometry of the map exp, and hence $\|v\|_\infty = d_\infty(f, h)$. Since $\varepsilon < \varepsilon_p$, and assuming $h \in U(f)$, the inverse $\exp_{f(t)}^{-1}$ is well defined at every point $h(t)$ and hence the equation displayed in the statement of the lemma is inverted by $v(t) = \exp_{f(t)}^{-1}(h(t))$. Finally, the maps $\exp_p(\cdot)$ and $\exp_p^{-1}(\cdot)$ depend differentiably on p, which implies h is an H_1-curve if and only if v is an H_1-vector field. Q.E.D.

8.15. Lemma: The 1–1 mapping η of Lemma 8.14, given by $\eta(h) = v$, is a C^{k-5} diffeormorphism between the open subset $U(f)$ of the manifold $H_1(S_1, M)$ and the open set $O(f)$ of the Hilbert space $H_1(S_1, TM_f)$.

8.16. Corollary: The mappings $\eta: U(f) \to O(f)$ of Lemma 8.14 and 8.15 define valid C^{k-5} coordinates on the manifold $H_1(S_1, M)$. We refer to them as standard coordinates near f.

Remark: It is possible to define the differentiable structure of $H_1(S_1, M)$ directly with the aid of Lemma 8.14 independently of Definition 8.11; in this case one has to show that the change-of-coordinates maps $\eta_2\eta_1^{-1}$ are of class C^{k-5}.

The proof of Lemma 8.15 is based on Theorem 8.10 and Whitney's embedding theorem. We start with some considerations which we shall need again.

8.17. Using Whitney's embedding theorem, we may take M as a C^k-submanifold of some R^N. The Whitney sum $TM \oplus \nu M$ of the tangent bundle TM and the normal bundle νM is then the trivial bundle $M \times R^N$. Using $M \subset R^N$ we embed the trivial bundle $M \times R^N$ in $R^N \times R^N$. Since the tangent bundle TM is a C^{k-1}-subbundle of the trivial bundle we thus get TM as a C^{k-1}-submanifold of $R^N \times R^N$. This embedding has the following properties. If we identify M with the zero section of TM, then this submanifold of TM is embedded in $R^N \times \{0\}$. Moreover the tangent space M_p of M and p is embedded as a linear subspace of $\{p\} \times R^N$ in such a way that the linear structure of M_p is preserved. In view of Theorem 8.10 we have then $H_1(S_1, TM)$ as a C^{k-5}-submanifold of $H_1(S_1, R^N \times R^N)$, and for fixed $f \in H_1(S_1, M)$, $H_1(S_1, TM_f)$ is a *linear* subspace (and therefore as a C^∞ submanifold) of $H_1(S_1, R^N \times R^N)$ and consequently $H_1(S_1, TM_f)$ is a C^{k-5}-submanifold of $H_1(S_1, TM)$.

In the same way the Whitney sum $TM \oplus TM$ is C^{k-1}-submanifold of $R^N \times R^N \times R^N$ so that the linear structure of the fibers is preserved. Consequently $H_1(S_1, TM \oplus TM)$ is a C^{k-5}-submanifold of $H_1(S_1, R^N \times R^N \times R^N)$ (Theorem 8.10) and $H_1(S_1, TM_f) \times H_1(S_1, TM_f)$ is a linear subspace of $H_1(S_1, R^N \times R^N \times R^N)$ and a C^{k-5}-submanifold of $H_1(S_1, TM \oplus TM)$.

Proof of Lemma 8.15: We assume 8.17. Consider the map $\Phi : TM \to M$ defined by $\Phi(v) = \exp_{p(v)}(v)$ (where $p(v)$ is the base point of v). Then $\Phi \in C^{k-2}$ and by Theorem 8.10 the induced map $\bar{\Phi}$: $H_1(S_1, TM) \to H_1(S_1, M)$ belongs to C^{k-5} (not C^{k-4} since TM is only C^{k-1}). But the restriction of $\bar{\Phi}$ to the C^{k-5}-submanifold $H_1(S_1, TM_f)$ (cf. 8.17) is the map η^{-1} of our Lemma, proving that $\eta^{-1} \in C^{k-5}$.

To show that we also have $\eta \in C^{k-5}$, we argue as follows. On the open subset $U = \{(p, q) \in M \times M; d(p, q) < \varepsilon_p\}$ the map γ: $U \to TM$ given by $\gamma(p, q) = \exp_p^{-1}(q)$ is well defined and C^{k-2}. By Theorem 8.10 γ induces a C^{k-5} map $\bar{\gamma}$ of an open subset of $H_1(S_1, M) \times H_1(S_1, M)$ into $H_1(S_1, TM)$. The domain of $\bar{\gamma}$ contains the C^{k-5} submanifold $\{f\} \times U(f)$ (cf. 8.13) and the restriction of $\bar{\gamma}$ to $\{f\} \times U(f)$ coincides with η by the proof of Lemma 8.14. Q.E.D.

The next Lemma shows that the Hilbert manifold $H_1(S_1, TM)$ may be identified with the tangent manifold $TH_1(S_1, M)$.

8.18. Lemma: Let $\Phi: TM \to M$ be the map given by $\Phi(v) = \exp(v)$, and let $\bar{\Phi}$ be the induced map (Theorem 8.10) of $H_1(S_1, TM) \to H_1(S_1, M)$. Let $\hat{\Phi}$ be the map defined by $\hat{\Phi}(v) = \frac{d}{ds}\bar{\Phi}(sv)\Big|_{s=0}$, so that $\hat{\Phi}: H_1(S_1, TM) \to TH_1(S_1, M)$. Then $\hat{\Phi}$ is a C^{k-6} diffeomorphism of $H_1(S_1, TM)$ onto $TH_1(S_1, M)$.

Proof: From the proof of Lemma 8.15 we have that $\bar{\Phi} \in C^{k-5}$ so that $\hat{\Phi} \in C^{k-6}$. Write $p(v)$ for the base point of the vector $v \in TM$, or for the base point curve of the H_1-vector field $v \in H_1(S_1, TM)$, as the case may be. Let $v \in H_1(S_1, TM)$. The point $\bar{\Phi}(v) \in H_1(S_1, M)$ has by Corollary 8.16 the coordinate v in the standard coordinate neighborhood $U(p(v))$ near $p(v)$. Thus $\frac{d}{ds}\bar{\Phi}(sv)\Big|_{s=0} = \hat{\Phi}(v)$ is that tangent vector of $H_1(S_1, M)$ at $p(v)$, which, in the coordinate system of $TH_1(S_1, M)$ corresponding to $U(p(v))$, has the coordinate $(0, v)$. This shows that $\hat{\Phi}$ is a 1–1 C^{k-6} mapping of $H_1(S_1, TM)$ onto $TH_1(S_1, M)$.

To prove that $\hat{\Phi}^{-1}$ is also of class C^{k-6} consider the coordinate system $\hat{U}(v)$ of $TH_1(S_1, M)$ corresponding to $U(p(v))$ and denote the C^{k-6}-coordinate map by $\hat{\eta}$, i.e. $\hat{\eta}: \hat{U}(v) \to O(p(v)) \times H_1(S_1, TM_{p(v)})$. We shall find a C^{k-5} map $\bar{\sigma}$ such that $\hat{\Phi}^{-1} = \bar{\sigma} \circ \hat{\eta}$, which proves $\hat{\Phi}^{-1} \in C^{k-6}$.

Now use 8.17. Let $p = p(v_1)$ and $q = \exp_p(v_1)$. By Lemma 8.1 there exists $\varepsilon_p > 0$ such that if $v_1, v_2 \in M_p$ and $|v_2| < \varepsilon_p$ then $\varrho(v_1, v_2) = \exp_q^{-1}(\exp_p(v_1 + v_2))$ defines a C^{k-2} function whose value is a vector tangent to M at q. Thus $\sigma(v_1, v_2) = \frac{d}{ds}\,\varrho(v_1, sv_2)\Big|_{s=0}$ defines a C^{k-3} map $\sigma: TM \oplus TM \to TM$. By Theorem 8.10 σ induces a C^{k-5} map $\bar{\sigma}: H_1(S_1, TM \oplus TM) \to H_1(S_1, TM)$ and therefore by restriction a C^{k-5} map (cf. 8.17)

$$\bar{\sigma}: H_1(S_1, TM_f) \times H_1(S_1, TM_f) \to H_1(S_1, TM).$$

For $(v_1, v_2) \in H_1(S_1, TM_f) \times H_1(S_1, TM_f)$ and $h = \bar{\Phi}(v_1)$ we have

$$\frac{d}{ds}(\exp_h^{-1}(\exp_f(v_1 + sv_2)))\Big|_{s=0} = \bar{\sigma}(v_1, v_2)$$

(cf. 8.10 (ii), more explicitly we have $R(s, v_1, v_2) = \varrho(v_1, sv_2)$ and $\sigma(v_1, v_2) = dR_{(0, v_1, v_2)}(1, 0, 0)$. Apply 8.10 (ii) to R and obtain

$$\bar{\sigma}(v_1, v_2)(t) = dR_{(0, v_1(t), v_2(t))}(1, 0, 0)$$
$$= \frac{d}{ds}\left(\exp_{h(t)}^{-1}(\exp_{f(t)}(v_1(t) + sv_2(t)))\Big|_{s=0}\right).$$

Thus

$$\exp_h^{-1}(\exp_f(v_1 + sv_2)) = s \cdot \bar{\sigma}(v_1, v_2) + o(s) \in H_1(S_1, TM_h).$$

Since $\bar{\Phi}(\exp_h^{-1}(\exp_f(v_1 + sv_2))) = \bar{\Phi}(v_1 + sv_2)$ and since $\eta^{-1} = \bar{\Phi}\Big|_{H_1(S_1, TM_f)}$ (by the proof of 8.15) it follows that

$$\frac{d}{ds}\bar{\Phi}(s\bar{\sigma}(v_1, v_2))\Big|_{s=0} = \frac{d}{ds}\bar{\Phi}(v_1 + sv_2)\Big|_{s=0},$$

i.e. $\hat{\Phi}(\bar{\sigma}(v_1, v_2))$ is that tangent vector of $H_1(S_1, M)$ at the point $h = \bar{\Phi}(v_1)$ whose coordinates in the coordinate neighborhood $\hat{U}(v_1)$ of $TH_1(S_1, M)$ are (v_1, v_2). In other words $\hat{\eta}(\hat{\Phi}(\bar{\sigma}(v_1, v_2))) = (v_1, v_2)$ or $\bar{\sigma} \circ \hat{\eta} = \hat{\Phi}^{-1}$, which completes the proof. Q.E.D.

Our next aim is to prove the differentiability of the energy integral and to introduce the intrinsic Riemannian metric for $H_1(S_1, M)$, more precisely:

8.19. Theorem: The energy J is a C^{k-5} function on $H_1(S_1, M)$.

8.20. Theorem: Suppose that we represent the tangent space to $H_1(S_1, M)$ at f by $H_1(S_1, TM_f)$ (cf. Corollary 8.16) and take as scalar product in the tangent space the intrinsic scalar product for $H_1(S_1, TM_f)$ defined in 8.6 (3). Then this scalar product defines a C^{k-6} Riemannian metric for $H_1(S_1, M)$.

Before we prove these two theorems, we make the following observations.

8.21. Let M be embedded in R^N and TM in $R^N \times R^N$ as described in 8.17. The C^{k-1} Riemannian metric g of M can be extended to a C^{k-2} Riemannian metric of R^N in such a way as to make M a totally geodesic submanifold (using the normal bundle of M in R^N and partitions of unity). This implies that the covariant derivative along a curve $f \in H_1(S_1, M)$ is the same if f is considered as a curve in M or as a curve in R^N. We write the scalar product as $g(p)(v, w)$ for (p, v) and $(p, w) \in R^N \times R^N$ (this is of course bilinear in v, w). The following statements are very similar to Lemma 6.9 and are proved in the same way.

(1) Let $b(p)(v, w)$ be bilinear in $v, w \in R^N$ and of class C^k in $p \in R^N$, and define a C^{k-2} function

$$\bar{b}: H_1(S_1, R^N) \times H_1(S_1, R^N) \times H_1(S_1, R^N) \to R$$

by

$$\bar{b}(f)(v, w) = \int_0^1 b(f(t))(\dot{v}(t), \dot{w}(t))\, dt.$$

(2) Let $b(\ )(\ ,\ ): R^N \to L^2(R^M)$ be a C^k map from R^N into the bilinear forms on R^M. Then we define a C^{k-2} map

$$\hat{b}(\ )(\ ,\ ): H_1(S_1, R^N) \to L^2(H_1(S_1, R^M))$$

by

$$\hat{b}(f)(v, w) = \int_0^1 b(f)(t))(\dot{v}(t), \dot{w}(t))\, dt.$$

(3) Let $c(\ ,\ )(\ ,\ ): R^N \times R^N \to L^2(R^M)$ be a C^k map from $R^N \times R^N$, which is linear in the second argument, into $L^2(R^M)$. Then we define a C^{k-1} map

$$\hat{c}(\ ,\ )(\ ,\ ): H_1(S_1, R^N) \times H_1(S_1, R^N) \to L^2(H_1(S_1, R^M))$$

by

$$\hat{c}(f, h)(v, w) = \int_0^1 c(f(t), \dot{h}(t))\,(v(t), \dot{w}(t))\, dt.$$

We show $c(f, h)\, ö\, L^2(H_1(S_1, R^M))$ to indicate the kind of changes which have to be made to adapt the proof of Lemma 6.9. We use Lemma 8.8 and Schwarz' inequality and we denite by e_i the i-th unit vector in R so that $h(t) = \Sigma\, e_i h_i(t)$.

$$|c(f(t), \dot{h}(t))\,(v(t), \dot{w}(t))| \leq \|c(f(t), \dot{h}(t))\|_{L^2(R^M)} \cdot |v(t)|_{R^M} \cdot |\dot{w}(t)|_{R^M}$$

$$\leq \Sigma\, |\dot{h}_i(t)| \cdot \|c(f(t), e_i)\|_{L^2(R^M)} \cdot \max_{t \in [0,1]} |v(t)| \cdot |\dot{w}(t)|$$

and therefore

$$|\hat{c}(f, h)(v, w)| \leq \max_{t \in [0,1} (\Sigma\, \|c(f(t), e_i)\|^2_{L^2(R^M)})^{1/2} \cdot \|h\|_{H_1(S_1, R^N)} \cdot 2\, \|v\|_{H_1(S_1, R^M)}$$

$$\cdot\, \|w\|_{H_1(S_1, R^M)}.$$

Proof of Theorem 8.19: With the notations of 8.21 $2J(f) = \int_0^1 g(f(t)) \times (\dot{f}(t), \dot{f}(t))\, dt$ is the restriction of the C^{k-4} function $\bar{g}$ on $H_1(S_1, R^N) \times H_1(S_1, R^N) \times H_1(S_1, R^N)$ to the diagonal of the C^{k-5} submanifold

$$H_1(S_1, M) \times H_1(S_1, M) \times H_1(S_1, M).$$

Q.E.D.

Proof of 8.20: We identify $TH_1(S_1, M)$ and $H_1(S_1, TM)$ using Lemma 8.18, i.e. we represent tangent vectors of $H_1(S_1, M)$ by elements of $H_1(S_1, TM)$. For $v \in H_1(S_1, TM)$ we denote the covariant derivative along the base point curve $p(v)$ by Dv/dt (cf. 8.21). Then the scalar product according to 8.6 (3) is

$$\langle v, w\rangle = \int_0^1 \left\{ g(p(v(t)))(v(t), w(t)) + g(p(v(t)))\left(\frac{Dv}{dt}, \frac{Dw}{dt}\right)\right\} dt,$$

where v and w have the same base point curve $p(v)$. We have to show that this formula defines a C^{k-6} section from M into the positive-definite, symmetric, continuous bilinear forms on $TH_1(S_1, M) = H_1(S_1, TM)$, cf. Lemma 8.18. Taking, for example, Cartesian coordinates for R^N we may write the covariant derivative along f as

$$\frac{Dv^i}{dt} = \frac{dv^i}{dt} + \Gamma^i_{jk}(f(t))\, v^j(t)\, \dot f^k(t)) \left(i = 1, \ldots, N \text{ and } \sum_{j,k=1}^{N}\right),$$

where the Γ^i_{jk} are C^{k-3} functions defined on R^N (cf. 8.21). We write $\Gamma(f(t))\,[v(t), \dot f(t)]$ for the vector $\{\Gamma^i_{jk}(f(t))\, v^j(t)\, \dot f^k(t)\} \in R^N$.

By 8.21 (2) and (3) we may define a C^{k-5} map

$$F: H_1(S_1, R^N) \times H_1(S_1, R^N) \to L^2(H_1(S_1, R^N))$$

by

$$F(f, h)(v, w) = \int_0^1 \{g(f(t))(v(t), w(t)) + g(f(t))(\dot v(t) + \Gamma(f(t))\,[v(t), \dot h(t)],$$
$$\dot w(t) + \Gamma(f(t))\,[w(t), \dot h(t)])\}\, dt.$$

The restriction of F first to the diagonal of $H_1(S_1, R^{“}) \times H_1(S_1, R^{“})$, which we identify with $H_1(S_1, R^{“})$, and then to the submanifold $H_1(S_1, M)$, is again C^{k-5}. In other words, we have a C^{k-5} Riemannian metric on the trivial vector bundle $H_1(S_1, M) \times H_1(S_1, R^N)$ and consequently also a C^{k-5} Riemannian metric on the C^{k-5} subbundle $H_1(S_1, TM)$, cf. 8.17. We lose on more order of differentiability since the identification of $H_1(S_1, TM)$ with the tangent bundle $TH_1(S_1, M)$ is only C^{k-6} (cf. Lemma 8.18). This Riemannian metric induces the right topology by Theorem 8.9. Q.E.D.

We continue with some observations concerning a useful family of differentiable curves on $H_1(S_1, M)$.

8.22. Notation: An element $f \in H_1(S_1, M)$ will be called a *curve on* M or a *point of* $H_1(S_1, M)$ depending on the situation. A *curve* $\varkappa$ *on* $H_1(S_1, M)$ will always be a map $\varkappa: [a, b] \to H_1(S_1, M)$.

8.23. Definition: Let $f_0, f_1 \in H_1(S_1, M)$ be such that $d_\infty(f_0, f_1) < \varepsilon_p$, which implies that the shortest geodesics on M joining $f_0(t)$ and $f_1(t)$ are unique. Then put:

$$\gamma(s, t) = \exp_{f_0(t)}(s \cdot \exp^{-1}_{f_0(t)}(t)(f_1(t))).$$

From this function of two variables we obtain a curve $\gamma : s \to \gamma(s) \in H_1(S_1, M)$ by writing $\gamma(s)(t) = \gamma(s, t)$.

8.24. Lemma: The γ-curves of Definition 8.23 are C^{k-5} differentiable. The tangent vector $\dfrac{\partial \gamma}{\partial s}(0) \in H_1(S_1, TM_f)$ is the coordinate of f_1 in the standard coordinate system near f_0 (cf. Lemma 8.14 and Corollary 8.16). We have $\left|\dfrac{\partial \gamma}{\partial s}\right|(s, t) = d_M(f_0(t), f_1(t))$ (for all s, $0 \leqq s \leqq 1$); hence $\left\|\dfrac{\partial \gamma}{\partial s}\right\|_\infty = d_\infty(f_0, f_1)$.

Proof: In the coordinate system centered at f_0 the γ-curves have the following representation.

$$v(s)(t) = s \exp^{-1}_{f_0(t)}(f_1(t)) \quad \text{where} \quad v(s) \in H_1(S_1, TM_{f_0}).$$

This implies that the γ-curves are as often differentiable as the change of coordinates map, i.e. are C^{k-5}. Moreover, $\left.\dfrac{d}{ds} v(s)\right|_{s=0}$ is the coordinate of f_1. $\left|\dfrac{\partial \gamma}{\partial s}(s, t)\right|$ is the length of the tangent vector to the geodesic $\gamma(s, t)$, $t = \text{const.}$, and therefore equals $d_M(f_0(t), f_1(t))$.

8.25. Lemma: A C^1-curve $\varkappa : [0, 1] \to H_1(S_1, M)$ considered as map $[0, 1] \times [0, 1] \to M$ by putting $\varkappa(s, t) = \varkappa(s)(t)$ is a homotopy between the end points $\varkappa(0)$, $\varkappa(1) \in H_1(S_1, M)$. Moreover $\dfrac{\partial \varkappa}{\partial s}(s, t)$ is a continuous vector field on M along $\varkappa$. This implies that the deformation paths $\varkappa(s, t_0)$, $t_0 = $ constant, are rectifiable curves on M and that their length depends continuously on t.

Proof: Since J is continuous and $\varkappa[0, 1]$ is compact (in $H_1(S_1, M)$) we have $\max_{s \in [0,1]} J(\varkappa(s)) = A < \infty$. Therefore, by Lemma 8.3, the $\varkappa(s)$ are an equicontinuous family of curves on M. Equicontinuity in one variable and continuity in the other implies continuity in both variables. In this case $\varkappa(s)(t_0)$ is also equicontinuous in s for $t_0 \in [0, 1]$. To see this let $v(s)$ be the coordinate of $\varkappa(s)$ in some standard coordinate system on $H_1(S_1, M)$ (cf. 8.13 (2)), i.e.

for some $f \in H_1(S_1, M)$ and $\delta > 0$ let $\varkappa(s) \in U(f)$ and $v(s) \in H_1(S_1, TM_f)$ for $|s - s_0| < \delta$. By Lemma 8.8 we have

$$\|v(s) - v(s_0)\|_\infty^2 = \max_t |v(s)(t) - v(s_0)(t)|_{M_{f(t)}}^2$$
$$\leqq 2\langle v(s) - v(s_0), v(s) - v(s_0)\rangle,$$

and therefore the continuity of $v(s)$ implies the stated equicontinuity. To prove the second part of the Lemma observe that the derivative of a C^1-curve in $H_1(S_1, M)$ represents a tangent vector, so that $\frac{\partial \varkappa}{\partial s} \in H_1(S_1, TM_{\kappa(s)})$. Hence $\frac{\partial \varkappa}{\partial s}(s, t)$ is continuous in t for fixed s. Using coordinates we see as before that $\frac{\partial \varkappa}{\partial s}(s, t)$ is equicontinuous in s for fixed t_0, $t_0 \in [0, 1]$. This implies continuity of $\frac{\partial \varkappa}{\partial s}(s, t)$ in both variables.

Q.E.D.

8.26. Lemma: Let $s \to \varkappa(s)$ be a C^1-curve on $H_1(S_1, M)$ and $s \to w(s)$ be a C^1-vector field along $\varkappa(s)$. Define $\Phi(s, t) = \varkappa(s)(t)$ and $v(s, t) = w(s)(t)$. Then

$$\frac{D}{\partial t}\frac{\partial \phi}{\partial s} = \frac{D}{\partial s}\frac{\partial \phi}{\partial t} \quad \text{(almost everywhere)}$$

and

$$\left(\frac{D}{\partial t}\frac{D}{\partial s} - \frac{D}{\partial s}\frac{D}{\partial t}\right) v(s, t) = R\left(\frac{\partial \Phi}{\partial t}, \frac{\partial \Phi}{\partial s}\right) v(s, t) \quad \text{almost everywhere,}$$

where R is the C^{k-3} curvature tensor of M and $\frac{D}{\partial t}$ (resp. $D/\partial s$) is the covariant derivative along the curves $\Phi(s_0, t)$ (resp. $\Phi(s, t_0)$) on M.

Proof: The formulae are well known for Φ, $v \in C^2$ and extend by the usual limit arguments to the above situation.

Q.E.D.

8.27. Definition: Let $\varkappa : [0, 1] \to H_1(S_1, M)$ be a C^1-curve; then

$$L(\varkappa) = \int_0^1 \left\|\frac{\partial \varkappa}{\partial s}\right\| ds = \int_0^1 ds \left(\int_0^1 dt \left(\left(\frac{\partial \varkappa}{\partial s}, \frac{\partial \varkappa}{ds}\right) + \left(\frac{D}{\partial t}\frac{\partial \varkappa}{\partial t}, \frac{D}{\partial t}\frac{\partial \varkappa}{\partial s}\right)\right)\right)^{1/2}$$

is the *length* of $\varkappa$ and

$$E(\varkappa) = \frac{1}{2}\int_0^1 \left\| \frac{\partial \varkappa}{\partial s} \right\|^2 ds \text{ is the } \textit{energy} \text{ of } \varkappa.$$

8.28. Remark: $L^2(\varkappa) \leqq 2E(\varkappa)$ may be proved along precisely the lines of the proof of Lemma 8.3.

8.29. Definition: In view of Lemma 8.25, define the Riemannian distance between f_0, f_1 by

$$d(f_0, f_1) = \begin{cases} \inf\limits_{\substack{\kappa(0)=f_0 \\ \kappa(1)=f_1}} L(\varkappa) & \text{if } f_0 \text{ and } f_1 \text{ are homotopic as curves on } M \\ \infty & \text{if } f_0, f_1 \text{ are not homotopic.} \end{cases}$$

8.30. Theorem: $\left|\sqrt{2J(f_0)} - \sqrt{2J(f_1)}\right| \leqq d(f_0, f_1)$.

Proof: Either $d(f_0, f_1) = \infty$ and nothing has to be proved or there is a C^1-curve $\varkappa$ joining f_0 and f_1. In this latter case we have

$$J(\varkappa(s)) = \frac{1}{2}\int_0^1 \left(\frac{\partial \varkappa}{\partial t}(s,t), \frac{\partial \varkappa}{\partial t}(s,t) \right) dt$$

$$\frac{d}{ds}\sqrt{2J(\varkappa(s))} = \frac{1}{\sqrt{2J(\varkappa(s))}} \int_0^1 \left(\frac{D}{\partial s}\frac{\partial \varkappa}{dt}, \frac{\partial \varkappa}{\partial t} \right) dt,$$

Hence, using Lemma 8.26 to change the order of differentiation and noting the identity of Definition 8.27,

$$\frac{d}{ds}\sqrt{2J(\varkappa(s))} \leqq \frac{1}{\sqrt{2J(\varkappa(s))}} \left(\int_0^1 \left| \frac{\partial \varkappa}{\partial t}(s,t) \right|^2 dt \right)^{1/2} \left(\int_0^1 \left| \frac{D}{\partial t}\frac{\partial \varkappa}{\partial s} \right|^2 \partial t \right)^{1/2}$$
$$\leqq \left\| \frac{\partial \varkappa}{\partial s} \right\|.$$

Therefore $\sqrt{2J(\varkappa(1))} - \sqrt{2J(\varkappa(0))} \leqq L(\varkappa)$. It follows by symmetry that $\left|\sqrt{2J(\varkappa(1))} - \sqrt{2J(\varkappa(0))}\right| \leqq L(\varkappa)$, and taking infima over $\varkappa$ we obtain the theorem.

Q.E.D.

The following theorem is a generalization of Lemma 8.8.

8.31. Theorem: $d_\infty^2(f_0, f_1) \leqq 2d^2(f_0, f_1)$.

Proof: We must show $d_\infty^2(f_0, f_1) \leqq L^2(\varkappa)$ for any C^1-curve joining f_0 and f_1, cf. Definition 8.13 (1) and 8.29. Let $t_m \in [0, 1]$ be such that $d_\infty(f_0, f_1) = d_M(f_0(t_m), f_1(t_m))$. Then by Lemma 8.25 $\varkappa(s, t_m)$ is a C^1-curve on M and we have

$$d_\infty^2(f_0, f_1) = d_M^2(f_0(t_m), f_1(t_m)) \leqq \left(\int_0^1 \left|\frac{\partial\varkappa}{\partial s}(s, t_m)\right| ds\right)^2$$

$$\leqq \int_0^1 \max_t \left|\frac{\partial\varkappa}{\partial s}(s, t)\right| ds\Bigg)^2 \leqq 2\left(\int_0^1 \left\|\frac{\partial\varkappa}{\partial s}\right\| ds\right)^2 = 2L^2(\varkappa)$$

by Lemma 8.8.
Q.E.D.

8.32. Theorem: $H_1(S_1, M)$ is a complete metric space.

Proof: Let $\{f_n\}$ be a Cauchy sequence in $H_1(S_1, M)$. By Theorem 8.31 $\{f_n\}$ is a d_∞-Cauchy sequence, i.e. f_n converges uniformly to a continuous curve f on M. Since the coordinate neighborhoods on $H_1(S_1, M)$ are defined using the d_∞-distance (cf. Definition 8.13 (2)) it follows that for large n the f_n and f all belong to some single coordinate neighborhood. The coordinates v_n of the curves f_n form a Cauchy sequence in the corresponding coordinate Hilbert space; this Cauchy sequence converges to some limit v. The point of $H_1(S_1, M)$ with the coordinate v coincides as continuous curve on M with f; this proves that $f_n \to f \in H_1(S_1, M)$ in the topology of $H_1(S_1, M)$. Q.E.D.

8.33. Theorem: If we consider each point of M to define a closed, constant curve, then M is embedded isometrically as a totally geodesic closed submanifold $\hat{M}$ of $H_1(S_1, M)$.

Proof: Every "constant" curve $\hat{p}: S_1 \to p \in M$ is determined by its image point p on M. The set

$$\hat{M} = J^{-1}(0) = \{f | f: S_1 \to M \text{ is a constant curve}\}$$

is a closed subset of $H_1(S_1, M)$ since $J(f) = 0$ if and only if f is constant Let $U(\hat{p})$ be the standard neighborhood (cf. Corollary 8.16) of the constant curve $\hat{p}$ in $H_1(S_1, M)$. Then $f \in U(\hat{p}) \cap \hat{M}$ if and only if the coordinate of f is a constant vector field along $\hat{p}$. The constant vector fields form an n-($=\dim M$) dimensional (hence closed) subspace of the coordinate space $H_1(S_1, TM_{\hat{p}})$ and therefore $\hat{M}$ is a closed submanifold of $H_1(S_1, M)$. Next we show that if two constant curves $\hat{p}_0, \hat{p}_1$ are joined in $H_1(S_1, M)$ by a C^1-curve $\varkappa$, then they can be joined by a curve $\hat{\varkappa}: I \to \hat{M}$ such that $L(\hat{\varkappa}) \leqq L(\varkappa)$. Indeed, by Lemma 8.25, the curve $\varkappa_\tau: I \to M$, defined by $\varkappa_\tau(s) = \varkappa(s, \tau)$ is a C^1-curve on M. Using any such curve, we may define a curve $\hat{\varkappa}_\tau: I \to \hat{M}$ on $\hat{M}$ by putting $\hat{\varkappa}_\tau(s)(t) = \varkappa(s, \tau)$, $0 \leqq t \leqq 1$. We have

$$L(\hat{\varkappa}_\tau) = \int_0^1 ds \left\| \frac{\partial \hat{\varkappa}_\tau}{\partial s} \right\| = \int_0^1 ds \left\{ \int_0^1 dt \left(\left| \frac{\partial \hat{\varkappa}_\tau}{\partial s}(s)(t) \right|^2 + \left| \frac{D}{\partial t} \frac{\partial \hat{\varkappa}_\tau}{\partial s} \right|^2 \right) \right\}^{1/2}$$

$$= \int_0^1 ds \left| \frac{\partial \varkappa}{\partial s}(s, \tau) \right| = L_M(\varkappa_\tau),$$

since $\partial \hat{\varkappa}_\tau / \partial s$ is independent of t and $\dfrac{D}{dt} \dfrac{\partial \hat{\varkappa}_\tau}{\partial s} = 0$. Therefore the curve $\varkappa_\tau$ on M and the curve $\hat{\varkappa}_\tau$ on $\hat{M}$ have the same length. Moreover

$$\inf_\tau L_M(\varkappa_\tau) = \inf_\tau \int_0^1 ds \left| \frac{\partial \varkappa}{\partial s}(s, \tau) \right| \leqq \int_0^1 \int_0^1 d\tau\, ds \left| \frac{\partial \varkappa}{\partial s}(s, \tau) \right|$$

$$\leqq \int_0^1 ds \left(\int_0^1 d\tau \left| \frac{\partial \varkappa}{\partial s}(s, \tau) \right|^2 \right)^{1/2}$$

by the Cauchy-Schwarz inequality, so that

$$\inf_\tau L_M(\varkappa_\tau) \leqq \int_0^1 ds \left\| \frac{\partial \varkappa}{\partial s} \right\| = L(\varkappa).$$

By Lemma 8.25, $L_M(\varkappa_\tau)$ depends continuously on τ, hence there is a τ^* for which the above infimum is assumed. Putting $\hat{\varkappa} = \hat{\varkappa}_{\tau^*}$ we have $L_M(\varkappa_{\tau^*}) = L(\hat{\varkappa}) \leqq L(\varkappa)$. From this it follows that the shortest geodesic joining p_0 and p_1 on M, if considered as a curve on $\hat{M}$ joining $\hat{p}_0$ and $\hat{p}_1$, is also the shortest curve connecting $\hat{p}_0$ and $\hat{p}_1$ in $H_1(S_1, M)$, and is therefore a geodesic in $H_1(S_1, M)$. Thus plainly $d_M(p_0, p_1) = d(\hat{p}_0, \hat{p}_1)$ so that $\hat{M} \subseteq H_1(S_1, M)$ is totally geodesic.

Q.E.D.

We proceed to a more detailed discussion of the γ-curves. These γ-curves are "short enough" to eliminate the need to refer explicitly to geodesics on $H_1(S_1, M)$ in some situations with which we shall deal below.

For use in the following proofs we recall some facts which have been established above.

$$\left\|\frac{\partial\gamma}{\partial s}(s)\right\|_\infty = d_\infty(f_0, f_1) \leqq \sqrt{2}d(f_0, f_1) \leqq \sqrt{2}L(\gamma) \leqq 2\sqrt{E(\gamma)};$$

(cf. Lemmas 8.24, 8.31, 8.29, and 8.28).

8.34. Lemma: Let $\|R\|$ be a bound for the norm of the curvature tensor on M and let $\left\|\frac{\partial\gamma}{\partial s}\right\|_\infty = d_\infty$. Then we have

(1) $$\max_s \sqrt{2J(\gamma(s))} \leqq (1 - \|R\|\, d_\infty^2)^{-1}\left(\sqrt{2J(f_0)} + \left\|\frac{\partial\gamma}{\partial s}(0)\right\|\right);$$

(2) $$\max_s \left|\left\|\frac{\partial\gamma}{\partial s}(s)\right\| - \left\|\frac{\partial\gamma}{\partial s}(0)\right\|\right| \leqq 2\,\|R\|\, d_\infty^2\, s \max_{0<\sigma<1} \sqrt{2J(\gamma(\sigma))};$$

(3) $$\left|L(\gamma) - \left\|\frac{\partial\gamma}{\partial s}(0)\right\|\right| \leqq 2\,\|R\|\, d_\infty^2 \max_s \sqrt{2J(\gamma(s))};$$

(4) $$\left|\sqrt{2E(\gamma)} - \left\|\frac{\partial\gamma}{\partial s}(0)\right\|\right| \leqq 2\,\|R\|\, d_\infty^2 \max_s \sqrt{2J(\gamma(s))}.$$

Remarks: (1) gives a bound for the energy integral along a γ-curve in terms of the energy of the one end point, $J(f_0)$, and in terms of the *coordinate* of the other end point, $\frac{\partial\gamma}{\partial s}(0)$.

(2) shows that a γ-curve is parametrized almost proportionally to arc length.

Proof: (3) and (4) are immediate consequences of (2). To prove (2), we note that, by 8.6 (3) and 8.8 (4)

$$\frac{d}{ds}\left\|\frac{\partial\gamma}{\partial s}(s)\right\| = \frac{1}{\left\|\frac{\partial\gamma}{\partial s}(s)\right\|}\int_0^1 dt\left(\left(\frac{D}{\partial s}\frac{\partial\gamma}{\partial s}, \frac{\partial\gamma}{\partial s}\right) + \left(\frac{D}{\partial s}\frac{D}{\partial t}\frac{\partial\gamma}{\partial s}, \frac{D}{\partial t}\frac{\partial\gamma}{\partial s}\right)\right).$$

By integrating this equation and using $\frac{D}{\partial s}\frac{\partial\gamma}{\partial s} = 0$, 8.26, $\left|\frac{\partial\gamma}{\partial s}(s, t)\right| \leqq d_\infty$

and the Cauchy-Schwarz inequality we obtain

$$\left| \left\| \frac{\partial \gamma}{\partial s}(\sigma) \right\| - \left\| \frac{\partial \gamma}{\partial s}(0) \right\| \right|$$

$$\leqq \int_0^\sigma ds \left(\frac{1}{\left\| \frac{\partial \gamma}{\partial s}(s) \right\|} \int_0^1 dt \left| \left(R\left(\frac{\partial \gamma}{\partial s}, \frac{\partial \gamma}{\partial t} \right) \frac{\partial \gamma}{\partial s}, \frac{D}{\partial t} \frac{\partial \gamma}{\partial s} \right) \right| \right)$$

$$\leqq \|R\|\, d_\infty^2 \int_0^\sigma ds \left(\sqrt{2J(\gamma(s))}\, \frac{\left(\int_0^1 \left| \frac{D}{\partial t} \frac{\partial \gamma}{\partial s} \right|^2 dt \right)^{1/2}}{\left\| \frac{\partial \gamma}{\partial s}(s) \right\|} \right)$$

$$\leqq \|R\|\, d_\infty^2 \int_0^\sigma ds \sqrt{2J(\gamma(s))}\,.$$

This proves (2). From 8.30 we have

$$\left| \sqrt{2J(\gamma(\sigma))} - \sqrt{2J(f_0)} \right| \leqq \int_0^\sigma \left\| \frac{\partial \gamma}{\partial s}(s) \right\| ds.$$

Therefore (2) gives

$$\left| \sqrt{2J(\gamma(\sigma))} - \sqrt{2J(f_0)} \right| \leqq \left\| \frac{\partial \gamma}{\partial s}(0) \right\| + \|R\|\, d_\infty^2 \max_{0<s<1} \sqrt{2J(\gamma(s))},$$

which implies (1).

Q.E.D.

8.35. Theorem: The first derivative of the energy integral at $f \in H_1(S_1, M)$ is the continuous linear map $dJ_f : H_1(S_1, TM_f) \to R$ given by the formula

$$dJ_f(w) = \int_0^1 \left(\frac{D}{\partial t} w(t), \frac{\partial f}{\partial t} \right) dt, \quad w \in H_1(S_1, TM_f).$$

Moreover,

$$|dJ_f(w)| \leqq \sqrt{2J(f)} \cdot \|w\|.$$

Proof: Let $f_1 \in U(f_0)$ and let $\gamma(s)$ be the γ-curve joining f_0 and f_1. Then the coordinate of f_1 is $\frac{\partial \gamma}{\partial s}(0)$ (cf. 8.24) and we must prove that

$$\left| J(f_1) - J(f_0) - dJ_{f_0}\left(\frac{\partial \gamma}{\partial s}(0) \right) \right| = o\left(\left\| \frac{\partial \gamma}{\partial s}(0) \right\| \right).$$

Now, using 8.26, we see from 8.8 (4), from

$$J(f_1) - J(f_0) = \int_0^1 ds \frac{d}{ds}\left(\frac{1}{2}\int_0^1 \left(\frac{\partial\gamma}{\partial t}, \frac{\partial\gamma}{\partial t}\right) dt\right)$$

and from

$$\int_0^1 \left(\frac{D}{\partial s}\frac{\partial\gamma}{\partial t}(s,t), \frac{\partial\gamma}{\partial t}\right) dt\Big|_0^s = \int_0^s d\sigma \int_0^1 dt \left\{\left(\frac{D}{\partial s}\frac{\partial\gamma}{\partial t}, \frac{D}{\partial s}\frac{\partial\gamma}{\partial t}\right) + \left(\frac{D}{\partial s}\frac{D}{\partial s}\frac{\partial\gamma}{\partial t}, \frac{\partial\gamma}{\partial t}\right)\right\}$$

that

$$\left|J(f_1) - J(f_0) - \int_0^1 \left(\frac{D}{\partial t}\frac{\partial\gamma}{\partial s}(0), \frac{\partial f_0}{\partial t}\right) dt\right|$$

$$= \left|\int_0^1 ds \int_0^s d\sigma \int_0^1 dt \left\{\left(\frac{D}{\partial t}\frac{\partial\gamma}{\partial s}, \frac{D}{\partial t}\frac{\partial\gamma}{\partial s}\right) - \left(R\left(\frac{\partial\gamma}{\partial t}, \frac{\partial\gamma}{\partial s}\right)\frac{\partial\gamma}{\partial s}, \frac{\partial\gamma}{\partial t}\right)\right\}\right|$$

$$\leqq \int_0^1 ds \left\{\left\|\frac{\partial\gamma}{\partial s}(s)\right\|^2 + \left\|\frac{\partial\gamma}{\partial s}\right\|_\infty^2 \|R\|\, 2J(\gamma(s))\right\} \leqq C_1 \left\|\frac{\partial\gamma}{\partial s}(0)\right\|^2,$$

where $C_1 = C_1\left(\|R\|, J(f_0), \left\|\frac{\partial\gamma}{\partial s}(0)\right\|\right)$ by Lemma 8.34 (1) and 8.34 (2). Therefore $dJ_{f_0}\left(\frac{\partial\gamma}{\partial s}(0)\right) = \int_0^1 \left(\frac{D}{\partial t}\frac{\partial\gamma}{\partial s}(0), \frac{\partial f_0}{\partial t}\right) dt$, which proves our first conclusion. The inequality $|dJ_f(w)| \leqq \sqrt{2J(f)}\,\|w\|$ now follows from the Cauchy-Schwarz inequality.

Q.E.D.

Remark: If one carefully notes those parts of earlier results which have been used in the above proof, one finds that it not only establishes a formula for the derivative assuming differentiability, but actually proves the differentiability of J.

8.36. Definition: The continuous linear functional $dJ_f(w)$ on the Hilbert space $H_1(S_1, TH_f)$ can be represented as the earlier product of w and of a vector which we call grad $J(f)$. More specifically, we have (cf. 8.35)

$$dJ_f(w) = \langle \text{grad}\, J(f), w\rangle = \int_0^1 \left(\frac{D}{\partial t}w(t), \frac{\partial f}{\partial t}\right) dt$$

$$= \int_0^1 \left\{(\text{grad}\, J(f)(t), w(t)) + \left(\frac{D}{\partial t}\text{grad}\, J(f)(t), \frac{D}{\partial t}w(t)\right)\right\} dt.$$

8.37. Remark: From 8.35 it follows that $\|\text{grad } J(f)\| \leqq \sqrt{2J(f)}$, and from 8.36 that $\|\text{grad } J(f)\|^2 = \int_0^1 \left(\frac{D}{\partial t} \text{grad } J(f), \frac{\partial f}{\partial t}\right) dt$.

8.38. Theorem: The vector grad $J(f)$ tangent to $H_1(S_1, H)$ at f, may be interpreted as a vector field along the curve f on M, and is determined by the integral equation given in Definition 8.36. If f is smooth enough so that $\partial f/\partial t \in H_1(S_1, TM_f)$, then grad $J(f)(t)$ is the unique periodic solution of the differential equation

$$\frac{D^2}{\partial t^2} \text{grad } J(f)(t) - \text{grad } J(f)(t) = \frac{D}{\partial t}\frac{\partial f}{\partial t}.$$

Proof: Our first assertion is obvious. To prove the second, note that since $\frac{\partial f}{\partial t}$ and $w(t)$ are continuous, we have $\left(w(t), \frac{\partial f}{\partial t}\right)\Big|_0^1 = 0$. Hence integrating the left side of the integral equation of 8.36 by parts gives (use 8.8 (4))

$$-\int_0^1 \left(w(t), \frac{D}{\partial t}\frac{\partial}{\partial t} f(t)\right) dt = \int_0^1 \Big\{(\text{grad } J(f)(t), w(t)) + \left(\frac{D}{\partial t} \text{grad } J(f)(t), \frac{D}{\partial t} w(t)\right)\Big\} dt$$

This can hold for every $w \in H_1(S_1, TM_f)$ only if $D/\partial t$ grad $J(f)(t)$ is a continuous vector field along f, in which case integration of the last term by parts gives the desired result.

Q.E.D.

Recall that f is called a critical point of J if $\|\text{grad } J(f)\| = 0$.

8.39. Theorem: The critical points of J correspond precisely to the closed geodesics on M (including the constant curves, cf. 8.33).

Proof: Since the constant curves are of minimal energy, i.e., of energy zero, there is nothing to be proved for these curves. Next, let f be a non-constant closed geodesic and hence a C^2-curve. Then $\frac{\partial f}{\partial t} \in H_1(S_1, TM_f)$ and by Theorem 8.35 we have

$$dJ_f(w) = \int_0^1 \left(\frac{D}{\partial t} w(t), \frac{\partial f}{\partial t}\right) dt = -\int_0^1 \left(w(t), \frac{D}{\partial t}\frac{\partial f}{\partial t}\right) = 0,$$

since $\dfrac{D}{\partial t}\dfrac{\partial f}{\partial t} = 0$ for geodesics and $\left(w(t), \dfrac{\partial f}{\partial t}\right)\Big|_0^1 = 0$ by continuity. Therefore by 8.36 $\|\text{grad } J(f)\| = 0$ so that f is a critical point of J.

Next assume that f is any critical point of J, so that $\|\text{grad } J(f)\| = 0$. We shall prove $\dfrac{D}{\partial t}\dfrac{\partial f}{\partial t} = 0$ by constructing a vector field z which is parallel along f and which will turn out to be equal to $\dfrac{\partial f}{\partial t}$. (In what follows $D/\partial t$ will denote the covariant derivative along f.) To construct the vector field z, we first solve the equation $\dfrac{D}{\partial t} y = \dfrac{\partial f}{\partial t}$ putting $y(0) = 0$. Then y is a H_1-vector field along f, continuous everywhere except possibly for the fact that $y(0) \neq y(1)$. Next solve the differential equation $\dfrac{D}{\partial t} z = 0$ with the boundary condition $z(1) = +y(1)$. Again z is H_1 and continuous everywhere except possibly for the fact that $z(1) \neq z(0)$.

Note also that the solution of $\dfrac{D}{\partial t} x = z(t)$ with the end condition $x(1) = y(1)$ is $x(t) = tz(t)$.

Put $v(t) = x(t) - y(t)$. Then $v(t)$ is a H_1-vector field along f and $v(0) = v(1) = 0$. Hence $v \in H_1(S_1, TM_f)$. Since $\|\text{grad } J(f)\| = 0$ it follows by Theorem 8.35 that

$$dJ_f(v) = 0 = \int_0^1 \left(\frac{Dv}{\partial t}, \frac{\partial f}{\partial t}\right) dt.$$

Moreover

$$\int_0^1 \left(\frac{Dv}{\partial t}, z(t)\right) = (v(t), z(t))\Big|_0^1 - \int_0^1 \left(v(t), \frac{D}{\partial t} z\right) dt = 0.$$

Forming the difference of the last two equations and noting that $Dv/\partial t = Dx/\partial t - Dy/\partial t = z(t) - \partial f/\partial t$, we get

$$\int_0^1 \left(z(t) - \frac{\partial f}{\partial t}, z(t) - \frac{\partial f}{\partial t}\right) dt = 0.$$

This implies that $\partial f/\partial t = z(t)$ almost everywhere. Since $\partial f/\partial t$ is the derivative of an absolutely continuous function it follows that the indefinite integrals of $\partial f/\partial t$ and of $z(t)$ (in local coordinates) agree. But $z(t)$ is continuous and therefore everywhere equal to the derivative of f. Thus $\partial f/\partial t$ is continuous except possibly for a jump at $t = 0+, 1-$. Arguing in a similar way however we can show that $\partial f/\partial t$ is continuous except possibly for a jump at $t = \frac{1}{2}-$, $\frac{1}{2}+$. Thus $\partial f/\partial t$ is continuous everywhere and parallel along f, so that f is a closed geodesic.

Q.E.D.

8.40. Remark: We restate the Palais-Smale condition for the energy J for the convenience of the reader:

If $\{f_n\}$ is a sequence on $H_1(S_1, M)$ such that $J(f_n) \leqq A$ and $\|\text{grad } J(f_n)\|$ converges to 0, then $\{f_n\}$ has a subsequence which converges to a critical point.

8.41. Theorem: The Palais-Smale condition holds for J.

Proof: Since $H_1(S_1, M)$ is complete (by 8.32) and since $\|\text{grad } J(\cdot)\|$ is continuous it suffices to find a subsequence h_n of f_n which is a Cauchy sequence. Since by Lemma 8.3 the $\{f_n\}$ are an equicontinuous family on a compact manifold we can use Arzela's theorem to find a subsequence of $\{f_n\}$ which converges uniformly; suppose, without loss of generality, that $\{f_n\}$ has this property. We then have $d_\infty(f_n, f_m) < \varepsilon$ for $n, m \geqq N(\varepsilon)$. We now show that $\{f_n\}$ is a Cauchy sequence. The proof will rest on Lemma 8.34 and the following formula for γ-curves (cf. 8.23, 8.8 (4) and 8.26)

$$\begin{aligned}\frac{\partial J(\gamma(s))}{\partial s} &= \int_0^1 \left(\frac{D}{\partial t}\frac{\partial \gamma}{\partial s}, \frac{\partial \gamma}{\partial t}(s, t)\right) dt \\ &= \int_0^1 \left(\frac{D}{\partial t}\frac{\partial \gamma}{\partial s}(0, t), \frac{\partial \gamma}{\partial t}(0, t)\right) dt + \int_0^1\int_0^s \left|\frac{D}{\partial t}\frac{\partial \gamma}{\partial s}(s, t)\right|^2 dt\, ds \\ &\quad + \int_0^1\int_0^s \left(R\left(\frac{\partial \gamma}{\partial t}, \frac{\partial \gamma}{\partial t}\right)\frac{\partial \gamma}{\partial s}, \frac{\partial \gamma}{dt}\right) dt\, ds, \end{aligned} \tag{5}$$

Let γ_{nm} be the γ-curve (cf. 8.23) such that $\gamma_{nm}(0) = f_n$, $\gamma_{nm}(1) = f_m$. Then by 8.29 and 8.34, we have

$$\begin{aligned} d(f_n, f_m) &\leqq L(\gamma_{nm}) \\ &\leqq \left\|\frac{\partial \gamma_{nm}}{\partial s}(0)\right\| + \frac{2\,\|R\|\, d_\infty^2(f_n, f_m)}{1 - \|R\|\, d_\infty^2(f_n, f_m)}\left(\sqrt{2A} + \left\|\frac{\partial \gamma_{nm}}{\partial s}(0)\right\|\right). \end{aligned}$$

We now suppress the indices n and m and write $\partial\gamma/\partial s$ for $\partial\gamma_{nm}/\partial s$, d_∞ for $d_\infty(f_n, f_m)$ and ε for $\dfrac{2\|R\|d_\infty^2}{(1 - \|R\|d_\infty^2)^2}$.

Then the above formula yields

(6) $$d(f_n, f_m) \leqq L(\gamma) \leqq (1+\varepsilon)\left\|\frac{\partial\gamma}{\partial s}(0)\right\| + \varepsilon\sqrt{2A}.$$

Since

$$\left|\frac{dJ(\gamma(s))}{ds}\right| = \left|\int\left(\frac{D}{\partial t}\frac{\partial\gamma}{\partial s}, \frac{\partial\gamma}{\partial t}\right)dt\right| \leqq \|\text{grad}\, J(\gamma(s))\|\left\|\frac{\partial\gamma}{\partial s}\right\|$$

we obtain from (5)

$$\int_0^1\int_0^1\left|\frac{D}{\partial t}\frac{\partial\gamma}{\partial s}(s,t)\right|^2 dt\, ds \leqq \|\text{grad}\, J(\gamma(1))\|\left\|\frac{\partial\gamma}{\partial s}(1)\right\| + \|\text{grad}\, J(\gamma(0))\|\left\|\frac{\partial\gamma}{\partial s}(0)\right\| + \int_0^1\int_0^1\left|\left(R\left(\frac{\partial\gamma}{\partial s}, \frac{\partial\gamma}{\partial t}\right)\frac{\partial\gamma}{\partial s}, \frac{\partial\gamma}{\partial t}\right)\right| dt\, ds,$$

and using $\left\|\frac{\partial\gamma}{\partial s}\right\|_\infty = d_\infty$, 8.34 (2), $\varepsilon = \dfrac{2\,\|R\|\, d_\infty^2}{(1-\|R\|\, d_\infty^2)}$

and 8.34 (1)

(7) $$\int_0^1\int_0^1\left|\frac{D}{\partial t}\frac{\partial\gamma}{\partial s}(s,t)\right|^2 dt\, ds \leqq \|\text{grad}\, J(\gamma(1))\|\left(\left\|\frac{\partial\gamma}{\partial s}(0)\right\| + \varepsilon\left(\sqrt{2A} + \left\|\frac{\partial\gamma}{\partial s}(0)\right\|\right)\right) + \|\text{grad}\, J(\gamma(0))\|\left\|\frac{\partial\gamma}{\partial s}(0)\right\| + \varepsilon\left(A + \left\|\frac{\partial\gamma}{\partial s}(0)\right\|^2\right).$$

On the other hand by 8.27 we have

$$\int_0^1\int_0^1\left|\frac{D}{\partial t}\frac{\partial\gamma}{\partial s}(s,t)\right|^2 dt\, ds \geqq 2E(\gamma) - d_\infty^2$$

and, after multiplying 8.34 (4) by $2E(\gamma) + \left\|\frac{\partial\gamma}{\partial s}(0)\right\|$ and using 8.34 (1)

$$\int_0^1\int_0^1\left|\frac{D}{\partial t}\frac{\partial\gamma}{\partial s}(s,t)\right|^2 dt\, ds \geqq \left\|\frac{\partial\gamma}{\partial s}(0)\right\|^2 - \varepsilon\left(\sqrt{2A} + \left\|\frac{\partial\gamma}{\partial s}(0)\right\|\right)\left(\left\|\frac{\partial\gamma}{\partial s}(0)\right\| + \varepsilon\left(\sqrt{2A} + \left\|\frac{\partial\gamma}{\partial s}(0)\right\|\right)\right) - d_\infty^2.$$

Combining this inequality with (7) we obtain

$$\left\|\frac{\partial\gamma}{\partial s}(0)\right\|^2 (1 - \varepsilon - 2\varepsilon^2) \leqq d_\infty^2 + \varepsilon\sqrt{2A}\,(1+\varepsilon)\left\|\frac{\partial\gamma}{\partial s}(0)\right\| + (\varepsilon + 2\varepsilon^2)\,A$$

$$+ ((1+\varepsilon)\,\|\text{grad}\, J(\gamma(1))\| + \|\text{grad}\, J(\gamma(0))\|)$$

$$\left\|\frac{\partial\gamma}{\partial s}(0)\right\| + \varepsilon\sqrt{2A}\,\|\text{grad}\, J(\gamma(1))\|$$

or, since $\varepsilon = O(d_\infty^2)$

$$\left\|\frac{\partial\gamma}{\partial s}(0)\right\|^2 = O(d_\infty^2) + \left\|\frac{\partial\gamma}{\partial s}(0)\right\| O(d_\infty^2) + (\|\text{grad}\, J(\gamma(0))\| + \|\text{grad} J(\gamma(1))\|)\left\|\frac{\partial\gamma}{\partial s}(0)\right\|. \tag{8}$$

From (8) it follows first that $\frac{\partial\gamma_{nm}}{\partial s}(0)$ stays bounded as $n, m \to \infty$ and then since $d_\infty \to 0$ and $\|\text{grad}\, J(\gamma_{nm}(0))\| \to 0$, it also follows from (8) that $\left\|\frac{\partial\gamma_{nm}}{\partial s}(0)\right\| \to 0$ as $n, m \to \infty$. This and (6) completes the proof.

Q.E.D.

We may now draw various easy consequences of the Palais-Smale condition (for more details see Chapter IV).

8.42. Lemma: If J has no critical points in $J^{-1}([a, b])$ then there exist $\delta > 0$ and $\varepsilon > 0$ such that $\|\text{grad}\, J\| \geqq \varepsilon$ on $J^{-1}[(a - \delta, b + \delta])$.

Proof: Were this false, we could find a sequence $\{f_n\}$ such that $\lim J(f_n) \in [a, b]$ and $\lim \|\text{grad}\, J(f_n)\| = 0$. But then the Palais-Smale condition implies that J has a critical point in $J^{-1}([a, b])$.

Q.E.D.

8.43. Definition: As in Chapter IV, we define a vector field on $H_1(S_1, M)$ by $v(f) = -\text{grad}\, J(f)$. Integration of $\frac{\partial\Phi}{\partial s} = v(\phi)$ with the initial condition $\Phi(0, f) = f$ defines the "gradient deformation" $\Phi(s, f)$.

8.44. Lemma: We have $\frac{dJ(\Phi(s, f))}{ds} = -\|\text{grad}\, J(\Phi(s, f))\|^2$.

Proof: $\dfrac{dJ(\Phi(s,f))}{ds} = \langle \text{grad } J(\Phi), \dfrac{\partial \Phi}{\partial s} \rangle = -\|\text{grad } J(\Phi)\|^2$
(cf. 4.71).

Q.E.D.

Let the singular cycle (or curve) z be a representative of any nontrivial homology class (or homotopy class) H of the pair $(H_1(S_1, M), \hat{M})$ (cf. 8.33). The range (or carrier) of z is a compact subset of $H_1(S_1, M)$ on which J assumes a maximum. We make the following definition.

8.45. Definition:

$$c_H = \inf_{\{z\} \in H} \left(\max_{f \in \text{range } z} J(f) \right).$$

c_H is called the *critical value of H*.

8.46. Theorem: c_H is a critical level of J.

Proof: We assume $c_H > 0$ since 0 is a critical level of J. If c_H is not a critical level, then for some $\delta > 0$ and $\varepsilon > 0$ it follows by 8.42 that $\|\text{grad } J\| \geqq \varepsilon$ on $J^{-1}([c_H - \delta, c_H + \delta])$. By definition of c_H we can find $z \in H$ such that $\max_{f \in \text{range } z} J(f) \leqq c_H + \delta$. Now deform z using the gradient deformation for $0 \leqq s \leqq 2\delta/\varepsilon^2$. Then $\Phi\left(\dfrac{2\delta}{\varepsilon^2}, z\right) \in H$ and

$$\max_{f \in \text{range } \Phi(2\delta/\varepsilon^2, z)} J(f) \leqq c_H - \delta$$

by 8.44, contradicting the definition of c_H.

Q.E.D.

The preceding theorem does not by itself imply the existence of a single nontrivial closed geodesic since the possibility $c_H = 0$ is not excluded and since we do not know the existence of nontrivial homology or homotopy classes of $H_1(S_1, M), \hat{M})$. The remainder of our reasoning is aimed at overcoming this difficulty.

8.47. Theorem: There exists $e > 0$ such that $\hat{M}$ (cf. 8.33) is a deformation retract of $J^{-1}([0, e])$.

Remark: From 8.47 it follows $c_H > 0$ for any nontrivial homology (or homotopy) class H of $(H_1(S_1, M, M)$.

The proof of 8.47 will follow from Corollary 8.49.

8.48. Theorem: There exists $e > 0$ such that the flow lines of the gradient deformation (cf. 8.43) which start in $J^{-1}([0, e])$ have uniformly bounded length as $s \to \infty$, and such that each of these flow lines has a well defined limit point in $\hat{M}$ as $s \to \infty$.

8.49. Corollary: The uniform boundedness of the length of the flow lines which start in $J^{-1}([0, e])$ implies that for these flow lines the end point in $\hat{M}$ and the length depend continuously on the starting point. Consequently we can parametrize these flow lines proportional to arc length and thus get a retraction of $J^{-1}([0, e])$ to $\hat{M}$, proving Theorem 8.47.

Proof of Theorem 8.48: We shall find $e > 0$ such that for $f \in J^{-1}([0, e])$ we have

(1) $$\|\text{grad}\, J(f)\|^2 \geqq \tfrac{1}{5} J(f).$$

This estimate implies 8.48 in the following manner. By 8.44 the flow line starting at $\Phi(0)$ satisfies

$$\int_0^s \|\text{grad}\, J(\Phi(\sigma))\|^2\, d\sigma = J(\Phi(0)) - J(\Phi(s)) \leqq e.$$

Therefore $\|\text{grad}\, J(\Phi(s))\|$ is not bounded away from 0 on a flow line. By the Palais-Smale condition, there exists a sequence $\Phi(s_n)$ converging to a critical point. By (1) it follows that $\lim\limits_{n\to\infty} \Phi(s_n) \in \hat{M}$. Since $J(\Phi(s))$ is monotone decreasing this proves $\lim\limits_{s\to\infty} J(\Phi(s)) = 0$. Therefore, using 8.44 and (1), we obtain the uniform bound for the length:

(2) $$L(\Phi) = \int_0^\infty \left\| \frac{\partial \Phi}{\partial s} \right\| ds = \int_0^\infty \|\text{grad}\, J(\Phi(s))\|\, ds$$

$$= \int_0^\infty \frac{-\dfrac{dJ}{ds}\, ds}{\|\text{grad}\, J(\Phi(s))\|} \leqq \sqrt{5} \int_{Q=J(\Phi(0))}^{J(\Phi(0))} \frac{dJ}{\sqrt{J}}$$

$$= 2\sqrt{5}\sqrt{J(\Phi(0))} \leqq \sqrt{20 \cdot e}$$

If a flow line had two different limit points in $\hat{M}$, its length would have to be infinite. Thus (2) proves

$$\lim_{s\to\infty} \Phi(s) \in \hat{M},$$

as desired.

To complete the proof of 8.48, it only remains to prove (1). We do this by introducing local coordinates on M. Let ε_p, C, m_1, m_2 be as in Lemma 8.1

and in addition let ε_p be so small that $m_2 - m_1 \leqq m_1/64$. We define the quantity e above as

$$e = \min\left(2\varepsilon_p^2, \left(8C\left(\frac{n}{m_1}\right)^{3/2}\right)^{-2}\right) \qquad (n = \dim M).$$

By Corollary 8.4, every $f \in J^{-1}([0, e])$ satisfies $L_M(f) \leqq 2\varepsilon_p$ and therefore is completely contained in the domain of some geodesic parallel coordinate system. Since both sides of (1) are continuous it suffices to prove (1) for $f \in C^2$. Then by 8.38 grad $J(f)$ is the periodic solution of

$$\frac{D^2}{dt^2} y - y = \frac{D}{dt} \dot{f}. \tag{3}$$

(where here and in the following proof we write y for grad $J(f)$ and $\dot{f}$ for $\partial f/\partial t$).

Moreover, by the proof of 8.38, Dy/dt is absolutely continuous. Next define

$$z = \frac{Dy}{dt} - \dot{f}. \tag{4}$$

We find using (3) that

$$\frac{D}{dt} z = \frac{D^2 y}{dt^2} - \frac{D}{dt}\dot{f} = y \quad \text{and therefore} \quad \frac{D^2 z}{dt^2} - z = \dot{f}. \tag{5}$$

From 8.37 we have $\|y\|^2 = \int_0^1 \left(\frac{Dy}{dt}, \dot{f}\right) dt$ and hence, using (4), we get

$$\|y\|^2 = \int_0^1 (z + \dot{f}, \dot{f})\, dt = 2J(f) + \int_0^1 (z, \dot{f})\, dt. \tag{6}$$

We shall prove (1) by estimating $\int_0^1 (z, \dot{f})\, dt$ in (6) as follows. Let $f^i(t)$ and $z^i(t)$ be the coordinates of $f(t)$ and $z(t)$ in a parallel coordinate system whose domain contains f (see choice of ε_p above). Then

$$\begin{aligned} \left|\int_0^1 (z(t), \dot{f}(t))\, dt\right| \leqq & \left|\int_0^1 \delta_{ik} z^i(0)\, \dot{f}^k(t)\, dt\right| \\ & + \left|\int_0^1 \delta_{ik}\,(z^i(t) - z^i(0))\, \dot{f}^k(t)\, dt\right| \\ & + \left|\int_0^1 \{(z(t), \dot{f}(t)) - \delta_{ik} z^i(t)\, \dot{f}^k(t)\}\, dt\right|. \end{aligned} \tag{7}$$

Since f is a closed curve contained in a single coordinate system we have

$$\int_0^1 \delta_{ik} z^i(0) \dot{f}^k(t)\, dt = 0. \tag{8}$$

Before we estimate the next term observe that we have from (4) and (5) using 8.37 again

$$\|z\|^2 = \int_0^1 \left\{(z, z) + \left(\frac{Dz}{dt}, \frac{Dz}{dt}\right)\right\} dt \tag{9}$$

$$= \int_0^1 \left\{\left(\frac{Dy}{dt} - \dot{f}, \frac{Dy}{dt} - \dot{f}\right) + (y, y)\right\} dt = 2J(f) - \|y\|^2.$$

Lemma 8.2, 8.1 (1), the Cauchy-Schwarz inequality, and (9) give

$$|\textstyle\int \{(z(t), \dot{f}(t)) - \delta_{ik} z^i(t) \dot{f}^k(t)\}\, dt| \tag{10}$$

$$\leqq \int_0^1 16 \frac{m_2 - m_1}{m_1} |z(t)| \cdot |\dot{f}(t)|\, dt$$

$$\leqq 16 \frac{m_2 - m_1}{m_1} \|z\| \sqrt{2J(f)} \leqq 16 \frac{m_2 - m_1}{m_1} 2J(f).$$

To estimate the second term in (7) note that $\dfrac{Dz}{dt} = y$ (cf. (5)) is in coordinates equivalent to

$$\frac{dz^i}{dt}(t) = y^i(t) - \Gamma^i_{jl}(f(t))\, z^j(t)\, \dot{f}^l(t),$$

so that integration and 8.1 (1) give

$$|z^i(t) - z^i(0)| \leqq \left|\int_0^t y^i(t)\, dt\right| + C \int_0^t \sum_{j,l} |z^j(t)| \cdot |\dot{f}^l(t)|\, dt. \tag{11}$$

By the Cauchy-Schwarz inequality, 8.1 (1), and (9), we have

$$\int_0^t \sum_{j,l} |z^j(t)|\, |\dot{f}^l(t)|\, dt \leqq n \left(\int_0^1 \sum_j |z^j|^2\, dt\right)^{1/2} \left(\int_0^1 \sum_l |\dot{f}^l|^2\, dt\right)^{1/2} \tag{12}$$

$$\leqq \frac{n}{m_1} \|z\| \sqrt{2J(f)} \leqq \frac{n}{m_1} 2J(f).$$

Since

$$\int_0^1 \sum_{i,k} |\delta_{ik} \dot{f}^k(t)|\, dt \leqq \sqrt{\frac{n}{m_1}} \int_0^1 (\dot{f}(t), f(t))\, dt$$

and

$$\int_0^1 \sum_{i,k} \delta_{ik} \left| \int_0^1 y^i(\tau)\, d\tau \right| |\dot{f}^k(t)\, dt|$$
$$\leqq \frac{1}{m_1} \left(\int_0^1 (y(\tau), y(\tau))\, d\tau \right)^{1/2} \left(\int_0^1 (\dot{f}(t), \dot{f}(t))\, dt \right)^{1/2}$$

we obtain from (11), and (12) using also $\|y\| \sqrt{2J(f)} \leqq \|y\|^2 + \frac{1}{2}J(f)$,

$$(13) \qquad \left| \int_0^1 \delta_{ik}\, (z^i(t) - z^i(0))\, \dot{f}^k(t)\, dt \right|$$
$$\leqq \frac{1}{m_1} (\|y\|^2 + \tfrac{1}{2}J(f)) + C \left(\frac{n}{m_1}\, 2J(f) \right)^{3/2}.$$

The estimates (8), (10) and (13) of the right side of (7) inserted in (6) give

$$(14)\ \|y\|^2 \geqq J(f)\,(2 - 32\frac{m_2 - m_1}{m_1} - \frac{1}{2m_1} - 2C\left(\frac{n}{m_1}\right)^{3/2} \sqrt{2e}) - \frac{\|y\|^2}{m_1},$$

so that $\|y\|^2 \geqq \frac{1}{5} J(f)$ by the choices made for ε_p and e.
Q.E.D.

8.50. Theorem: On every compact Riemannian manifold M of class C^k, $k \geqq 6$, there is at least one nontrivial closed geodesic.

Proof: If M is not simply connected, then by Theorem 8.46 and 8.47 there exists a closed geodesic in every nontrivial homotopy class of closed curves. If M is simply connected there is a first nonvanishing homotopy group $\pi_l(M)$, $l \geqq 2$. We claim that $\pi_{l-1}(H_1(S_1, M), \hat{M})$ is nontrivial. This together with Theorem 8.46 and the Remark following Theorem 8.47 implies the present theorem.

We will prove our claim for the case $M = S^2$, and indicate the modifications needed to treat the general situation thereafter. Consider the spheres S^2 in R^3 and a line l tangent at p to S^2. Take the tangent plane to S^2 at p and rotate it around l, through 180°, until it is again tangent. The intersections of the intermediate planes with S^2 form a family of circles. Parametrize the intermediate planes with a parameter s running from 0 to 1 and call the corresponding circles of intersection $c(s)$. Then $c(0)$ and $c(1)$ are constant curves (cf. 8.33). Parametrize each circle with a parameter t running from 0 to 1, so that $c(s)(0) = c(s)(1) = p$ (e.g. take t proportional to arc length on $c(s)$). Then $c: [0, 1] \to H_1(S_1, M)$ is a curve in $H_1(S_1, M)$ which represents an element of $\pi_1(H_1(S_1, M), \hat{M})$. We claim that this homotopy element is nontrivial.

To prove this, first consider the map $\tilde{c}: I \times I \to S^2 = M$ given by $\tilde{c}(s, t) = c(s)(t)$. Note that $\tilde{c}(s, 0) = \tilde{c}(s, 1)$ for each s and $\tilde{c}(0, t) = p = \tilde{c}(1, t)$ for all t. We shall make boundary identifications in $I \times I$ so that the square becomes a sphere and $\tilde{c}$ induces a map $c^*: S^2 \to S^2 = M$ such that c^* is homotopic to the identity and therefore homotopically nontrivial. This is done as follows. For each $s \neq 0,1$ identify the two points $(s, 0)$ and $(s, 1)$; for $s = 0$ identify all the points $(0, t)$ to a single point; and for $s = 1$ identify all the points $(1, t)$ to a single point.

Assume now that c represents a trivial element of $\pi_1(H_1(S_1, M), \hat{M})$. Then there is a deformation Φ_r of c in $H_1(S_1, M)$ which deforms c to a curve $d: [0, 1] \to \hat{M}$ and which leaves the end points of c fixed, i.e. $\Phi_r(0) = c(0) = \hat{p} = c(1) = \Phi_r(1)$ for all r. We can assume that Φ_r is a differentiable deformation. Using Lemma 8.25, we can interpret Φ_r as a homotopy of $\tilde{c}$ on M. Since $\Phi_r(s)(0) = \Phi_r(s)(1)$ for each r and s (since we deal with spaces of closed curves), and also since $\Phi_r(0)(t) = p = \Phi_r(1)(t)$ for all t, Φ_r gives a homotopy of c^*. Now note that the curves $\Phi_1(s) = d(s)$ are constant curves on M, i.e. $\Phi_1(s)(t)$ is independent of t, and note also that $d(0) = d(1) = \hat{p}$. We have thus constructed a homotopy from the nontrivial map $c^*: S^2 \to S^2 = M$ to a map $d^*: S^2 \to S^2 = M$ such that the image $d^*(S^2)$ lies on the closed curve in $S^2 = M$ which is given by $s \to \Phi_1(s)(0)$. Clearly d^* is homotopically trivial, a contradiction.

In regard to the general case, we remark only that starting with a nontrivial element of the first nonvanishing homotopy group $\pi_l(M)$, i.e. with a (differentiable) map $F: S^l \to M$, one may define an associated $(l-1)$-parameter family of circles on S^l such that the F-image of the family of circles represents a differentiable element of $\pi_{l-1}(H_1(S_1, M), \hat{M})$. If this element is trivial, one constructs, as above, a homotopy which deforms F to a map G which can be considered as an element of $\pi_{l-1}(M)$ and is therefore trivial by the choice of $\pi_l(M)$, a contradiction which proves our theorem in the general case. Q.E.D.

Comments on Further Developments of the Theory of Closed Geodesics

When one tries to prove the existence of more than one nontrivial closed geodesic, one runs into the disturbing fact that associated with each closed geodesic one automatically has a one-parameter family of geodesics arising from the various different starting points for the parametrization of a closed curve. (Note however that critical points of J are always parametrized proportional to arc length.) It is easy to prove that the action of $O(2)$ on $H_1(S_1, M)$, which for $\alpha \in O(2)$ is given by $f(t) \to f(t + \alpha)$ (t and α of course taken mod 1) is continuous. By identifying orbits we then get a new space $\Pi(M)$ (in Klingenberg's notation) in which a single nontrivial closed geodesic is represented by exactly one point. (Multiple coverings of a geodesic are not identified to a single point by this process.) The space $\Pi(M)$ is no longer a manifold, but since the Riemann scalar product of $H_1(S_1, M)$ and the energy function are compatible with the action of $O(2)$ (i.e., are equivariant under the action of $O(2)$) one can prove many statements concerning the space $\Pi(M)$ by "lifting" them to the manifold $H_1(S_1, M)$. For example, the gradient deformation in $H_1(S_1, M)$ induces an energy decreasing deformation in $\Pi(M)$, so that Theorem 8.47 also holds for $\Pi(M)$. This is remarkable since the result seems to be inaccessible via the classic techniques using broken geodesics. In a paper to appear in the journal *Topology*, W. Klingenberg gives a complete description of the Z_2-homology of both $H_1(S_1, M)$ and $\Pi(M)$ for $M = S^n$. His method of calculation also applies to the projective spaces and the other symmetric manifolds of rank 1. Using this information he obtains a number $g(n)$ of "algebraically different" nontrivial closed geodesics for the case of $M = S^n$; specifically $g(n) = 2n - s - 1$ where $0 \leqq s = n - 2^h < 2^h$. "Algebraically different" means that the energies of these $g(n)$ geodesics are the critical values (8.45) corresponding to $g(n)$ pairwise subordinated homology classes; here and below, we call a homology class α *subordinated* to a homology class β if there exists a cohomology class such that α can be written as a cap product $\alpha = \cap\, \beta$. Unfortunately the possibility cannot be excluded that all the geodesics obtained are multiple coverings of a single geodesic.

The same result was proved using different methods by S. L. Alber. (On periodicity problems in the calculus of variations in the large, Amer. Math. Soc. Transl. (2) 14 (1960).) A. I. Fet proved that on every compact manifold there are at least 3 algebraically different nontrivial closed geodesics. (On the algebraic number of closed extremals on a manifold, Dokl. Akad. Nauk SSSR (N. S.) 88 (1953), 619–621). Klingenberg obtains this result also.

No criterion is known which allows one to decide whether two algebraically different geodesics are also geometrically different (i.e. whether the underlying simple covered closed geodesics are different). However, Lusternik and Schnirelmann (Sur les problèmes de trois géodésiques fermées sur les surfaces de genre 0, C. R. Acad. Sci. Paris 189 (1929), 269–271) showed that on manifolds of the type of the 2-sphere there exist 3 closed geodesics without self intersections.

Fet proves the following result: If all closed geodesics on a compact manifold are nondegenerate as critical points of J then there are at least 2 prime geodesics (A periodic problem in the calculus of variations, Dokl. Akad. Nauk SSSR (N.S.) 160 (1965), 287–289). Alber and Klingenberg announced that under restrictions on the curvature of a manifold M $(\frac{1}{4} < \min_M K/\max_M K \leqq 1)$ one can prove without difficulty that the geodesics constructed from subordinated homology classes are geometrically different. Finally, Klingenberg announced that certain special closed geodesics constructed from subordinated homology classes turn out to be simple and without self intersection.

Index